T0348375

RESEARCH IN
ACCOUNTING REGULATION

Volume 9 • 1995

RESEARCH IN
ACCOUNTING REGULATION

Editors: **GARY JOHN PREVITS**
Weatherhead School of Management
Department of Accountancy
Case Western Reserve University

Associate Editors: **LARRY M. PARKER**
Weatherhead School of Management
Case Western Reserve University

ROBERT ESKEW
Krannert Graduate School of Business
Purdue University

VOLUME 9 • 1995

 **JAI PRESS INC.**

Greenwich, Connecticut   *London, England*

CONTENTS

LIST OF CONTRIBUTORS   ix

EDITORIAL BOARD   xiii

INVITED REFEREES FOR VOLUME 9   xv

MAIN PAPERS

AUDITOR REPORTING FOR BANKRUPT
COMPANIES: EVIDENCE ON THE IMPACT
OF SAS 59
    Van E. Johnson and Inder K. Khurana   3

THE EUROPEAN UNION: REGULATION MOVES
FINANCIAL REPORTING TOWARD COMPARABILITY
    Kathleen R. Bindon and Helen Gernon   23

SECURITY PRICE RESPONSE ASSOCIATED WITH
THE ACCOUNTING REGULATION OF PURCHASE
COMBINATIONS INCREASING GOODWILL
    Kathleen Blackburn Hethcox   49

THE EFFECT OF PEER REVIEW ON
AUDIT ECONOMICS
    Gary Giroux, Donald Deis, and Barry Bryan   63

REDUCING THE INCIDENCE OF FRAUDULENT
FINANCIAL REPORTING: EVALUATING THE
TREADWAY COMMISSION RECOMMENDATIONS
AND POTENTIAL LEGISLATION
    Jerry R. Strawser, John O'Shaughnessy, and Philip H. Siegel   83

v

AN EMPIRICAL ANALYSIS OF THE COMPARABILITY
OF DISCLOSURE TENDENCIES WITHIN AND
ACROSS INDUSTRIES: THE CASE OF HAZARDOUS
WASTE LAWSUITS
    Philip Little, Michael Muoghalu, and David Robison   105

PERSPECTIVES

PUBLIC ACCOUNTING IN AN OLDER SOCIETY:
SOME KEY PERSONNEL ISSUES
    Stephen E. Loeb   121

RESEARCH REPORTS

AIDING AND ABETTING AFTER CENTRAL
BANK OF DENVER
    Mark A. Segal   153

REFORMING ACCOUNTANTS' LIABILITY
TO THIRD PARTIES AND THE PUBLIC INTEREST
    Zhemin Wang   163

THE CHANGING PROFILE OF THE AICPA:
DEMOGRAPHICS OF A MATURING PROFESSION
    Stephen J. Young   181

AUDIT CONFLICT AND COST STANDARDS
IN THE DEFENSE INDUSTRY
    Norma C. Holter   199

ASSESSING THE UTILITY OF CONTINUING
PROFESSIONAL EDUCATION FOR
CERTIFIED PUBLIC ACCOUNTANTS
    Paul J. Streer, Ronald L. Clark, and Margaret E. Holt   211

BOOK REVIEWS

SETTING STANDARDS FOR FINANCIAL REPORTING:
FASB AND THE STRUGGLE FOR CONTROL
OF A CRITICAL PROCESS
By Robert Van Riper
Reviewed by Elliott L. Slocum 225

ACCOUNTING CERTIFICATION, EDUCATIONAL,
& RECIPROCITY REQUIREMENTS: AN
INTERNATIONAL GUIDE
By Jack Fay
Reviewed by Tonya K. Flesher 229

THE CONTINENTAL BANK JOURNAL OF
APPLIED CORPORATE FINANCE
By Stern Stewart Management Services, Inc.
Reviewed by Nandini Chandar 231

FINANCIAL REPORTING IN NORTH AMERICA:
HIGHLIGHTS OF A JOINT STUDY
By Financial Accounting Standards Board
Reviewed by Kevin Brown 235

REPORTING ON ENVIRONMENTAL PERFORMANCE
By The Canadian Institute of Chartered Accountants
Reviewed by Susan Frazier 239

EDITORS' NOTES

REPORTING REFORMS IN THE ACCOUNTING PROFESSION:
MARKET TIERS AND CHANGING TRADING RULES
MAKE REFORM A NECESSITY
Gary John Previts and Larry M. Parker 245

LIST OF CONTRIBUTORS

Kathleen R. Bindon

Culverhouse School of
   Accountancy
University of Alabama

Kevin Brown

Weatherhead School of Management
Case Western Reserve University

Barry Bryan

School of Accountancy
Auburn University

Nandini Chandar

Weatherhead School of Management
Case Western Reserve University

Ronald L. Clark

School of Accountancy
Auburn University

Donald Deis

Department of Accounting
Louisiana State University

Tonya K. Flesher

Department of Accounting
University of Mississippi

Susan Frazier

Deloitte & Touche
Cleveland, Ohio

Helen Gernon

Charles H. Lundquist School of
   Business
University of Oregon

Gary Giroux

Department of Accounting
Texas A&M University

Kathleen Blackburn Hethcox

Department of Accounting and
   Business Law
University of Tennessee

Margaret E. Holt   Department of Adult Education
                                     University of Georgia

Norma Holter   Department of Accounting
                                     Towson State University

Van E. Johnson   Department of Accountancy
                                     Northern Illinois University

Inder K. Khurana   School of Accountancy
                                     University of Missouri-Columbia

Philip Little   Department of Accounting
                                     Western Carolina University

Stephen E. Loeb   Department of Accounting
                                     University of Maryland

Michael Muoghalu   Department of Economics,
                                       Finance, and Banking
                                     Pittsburgh State University

John O'Shaughnessy   Department of Acccounting
                                     San Francisco State University

David Robison   Department of Economics
                                     LaSalle University

Mark A. Segal   Department of Accounting
                                     University of South Alabama

Philip H. Siegel   Department of Accounting
                                     University of Houston

Elliott L. Slocum   Department of Accounting
                                     Georgia State University

Jerry R. Strawser   Department of Accounting
                                     University of Houston

Paul J. Streer

Department of Accounting
University of Georgia

Zhemin Wang

Department of Accounting
North Dakota State University

Stephen J. Young

Weatherhead School of Management
Case Western Reserve University

xiii

Research in Accounting Regulation, Volume 9

INVITED REFEREES

Aaron Ames
Ernst & Young

Michael Barrett, Esq.
Private practice, Washington, D. C.

Ralph Benke
James Madison University

David Bowers
Case Western Reserve University

John Brozovsky
Virgina Polytechnical Institute
    and State College

Ivan Bull
University of Illinois

Chee Chow
San Diego State University

Edmund Coulson
Ernst & Young

Richard Dieter
Arthur Andersen & Co., SC

Eugene Flegm
General Motors, Retired

Julia Grant
Case Western Reserve University

William Hall
Arthur Andersen & Co., SC, Retired

David Hardison, Esq.
Fried, Frank, Harris, & Shriver

John Hill
Indiana University

Dan Guy
American Institute of
Certified Public Accountants

Gregory Jonas
Arthur Andersen & Co., SC, Retired

Alan Lord
University of Alabama

Alan Mayper
University of North Texas

Joan McGloshen
The Ohio Society of Cerified
Public Accountants

Paul B. W. Miller
University of Colorado at
Colorado Springs

A. Tom Nelson
University of Utah

Terrence O'Keefe
University of Oregon

Kurt Pany
Arizona State University

Karen Pincus
University of Southern California

Laura Pollard
The Ohio Society of Certified
Public Accountants

J. Clarke Price
The Ohio Society of Certified
Public Accountants

PART I

MAIN PAPERS

AUDITOR REPORTING FOR
BANKRUPT COMPANIES:
EVIDENCE ON
THE IMPACT OF SAS NO. 59

Van E. Johnson and Inder K. Khurana

ABSTRACT

Connor (1986) suggests that one of the primary sources of the public's dissatisfaction with the audit profession is the (public's) expectation that the auditor will warn them about impending business failure. Prior research has documented that less than one-half of the companies filing for bankruptcy received a going concern audit report in the year prior to bankruptcy. Carmichael and Pany (1993) suggest that these results may not bode well for the profession in today's litigious environment and question whether auditor performance on this dimension has improved subsequent to the passage of *SAS 59*. This research attempts to assess whether, subsequent to *SAS 59*,

Research in Accounting Regulation, Volume 9, pages 3-22.
Copyright © 1995 by JAI Press Inc.
All rights of reproduction in any form reserved.
ISBN: 1-55938-883-8

bankrupt companies were more likely to receive modified audit reports prior to the bankruptcy. Results of this study indicate that a larger proportion of bankrupt companies received modified audit reports after *SAS 59* became effective. Limitations and implications of the research findings are also discussed.

INTRODUCTION

The auditor's responsibility to evaluate and report on a client's ability to continue as a going concern has attracted the attention of regulators, the financial press, members of Congress, and academicians. While the public clearly expects the auditor to warn them about impending business failures, many within the profession believe that the auditor should provide assurances regarding the fairness of a client's financial statements, not its financial condition. In an attempt to narrow the gap between the public's expectation and auditors' own perceptions of auditor responsibilities, the Auditing Standards Board (ASB) issued nine "expectation gap" auditing standards in 1988. Of the nine standards, *Statement on Auditing Standards [SAS] No. 59*, "The Auditor's Consideration of an Entity's Ability to Continue as a Going Concern" (AICPA 1988b) has been described as the most controversial because it increased the auditor's responsibilities for evaluating and reporting on an entity's ability to continue as a going concern (Guy and Sullivan 1988; Kaplan and Pany 1992).

In 1991, the ASB called for research to evaluate the impact of the expectation gap audit standards (Holstrum 1991). Summarizing prior research that addressed auditor reporting for going concern uncertainties, Carmichael and Pany (1993) reported that generally less than one-half of all failed companies received a modified audit report in the period prior to bankruptcy. However, because the studies they reviewed were all based on auditor reporting prior to the issuance of *SAS 59*, Carmichael and Pany questioned whether "auditor performance on this dimension improved subsequent to the passage of *SAS 59*."

The purpose of this study is to address the research question raised by Carmichael and Pany. Samples of bankrupt companies were identified for the periods before and after the issuance of *SAS 59*. For each failed company, the last audit report issued prior to the bankruptcy filing was examined and classified as either modified (if the report mentioned uncertainties regarding the entity's ability to continue as a going concern) or unmodified. The audit reports received by bankrupt companies prior to the bankruptcy filing were compared over two periods: the period when *SAS 34* was the authoritative guidance and the period when *SAS 59* was the authoritative guidance. The results of the study indicate that a larger proportion of bankrupt companies received a going concern modification after *SAS 59* became effective.[1] The results of a logistic regression model also indicate that, after controlling for financial condition and size, bankrupt companies were more likely to receive a going concern modification subsequent to *SAS 59*.

The remainder of this paper consists of the following sections. The first section discusses the historical development of auditor responsibility for reporting on uncertainties, summarizes relevant research, and concludes with a statement of a testable hypothesis. The second section outlines the research approach and the data sources together with the basic test strategy used. The third section discusses the empirical results, implications, and limitations of the research, and the final section concludes the paper.

BACKGROUND AND HYPOTHESIS DEVELOPMENT

Historical Background

The profession first formally considered the impact of uncertainties (including going concern uncertainties) on the audit report in *Statement on Auditing Procedure No. 15* (AICPA 1942). This statement suggested that uncertainties could be so great that the auditor might be unable to state an opinion or that an exception to the audit opinion might be necessary (Kaplan and Pany 1992). The Securities and Exchange Commission subsequently issued *Accounting Series Release No. 90* "Certification of Income Statements" (SEC 1962) and the AICPA issued *Statement on Auditing Procedure No. 33* (AICPA 1963) which required that the term "subject to" be used to qualify the audit report when there was uncertainty regarding a matter of accounting significance. In 1974, the ASB issued *SAS 2* "Reports on Audited Financial Statements" (AICPA 1974) which explicitly addressed uncertainties regarding the continued existence of clients and attempted to provide financial statement characteristics important for the going concern decision. *SAS 2* also indicated that significant uncertainty about a client's continued existence would result in a "subject to" qualified audit report.

In 1978, the Cohen Commission (Commission on Auditors' Responsibilities 1978) recommended that the auditor's role regarding uncertainties should be limited to assessing and reporting on the adequacy of information disclosures. The evaluation of business risks facing the company was, in the Commission's view, beyond the scope of auditor responsibilities. The Commission did not think that auditors were in a better position than other outsiders to predict a company's survival or failure (Menon and Schwartz 1986).

In March of 1981, the ASB issued *SAS 34*, "The Auditor's Consideration When a Question Arises About an Entity's Continued Existence" (AICPA 1981). In *SAS 34*, the ASB accepted the premise that audit reports should be modified for going concern uncertainties and attempted to improve practice by providing guidance in the form of quantifiable and nonquantifiable factors to assist auditors in evaluating the going concern status of a company (Kaplan and Pany 1992; Menon and Schwartz 1986). Under *SAS 34*, an entity's continuation was assumed. Accordingly, auditors were required to consider an entity's ability to continue as a going concern only when "contrary" information was detected during the audit. If after assessing a client's going concern status, the auditor had *both* substantial doubt about the entity's continued existence and questions about the recovery of recorded asset values, then the auditor was required to issue either a "subject to" qualified audit report or a disclaimer. No modification of the audit report was required if the auditor had substantial doubt about the client's continued existence, but believed that the recorded asset values were recoverable.

Research Prior to *SAS 59*

Several prior studies have provided empirical evidence documenting the relationship between bankruptcy and going concern audit reports issued under *SAS 34* and earlier authoritative standards. Generally, the researchers in these studies identified a sample of companies that filed for bankruptcy within a chosen time frame by referring to some published source (usually the *Wall Street Journal Index*). The researchers then examined each failed company's audit report from the period prior to bankruptcy and classified the report as modified if the report was either a "subject to" qualification or a disclaimer duc to uncertainty regarding an entity's ability to continue as a going concern. The researcher classified the audit report as unmodified if the report was either unqualified or qualified for reasons other than a going concern uncertainty.

Prior studies that have examined the audit reports received by failed companies are summarized in Table 1. In one of the first studies, Altman and McGough (1974) identified 28 companies that had filed for bankruptcy between 1970 and 1973 and found that 13 of the 28 companies (46.4%) received a modified audit report in the period prior to bankruptcy. Altman (1982) updated his initial study by examining the audit reports of 81 companies that filed for bankruptcy between 1974 and 1982. He found that 39 of the companies (48.1%) received a modified audit report in the period prior to the bankruptcy.

Menon and Schwartz (1986) identified 147 companies listed on the New York or American Stock Exchanges that filed for bankruptcy between 1974 and 1983. Their examination of the last audit report issued prior to bankruptcy indicated that 63 of the companies (42.9%) received modified audit reports due to going concern uncertainties. McKeown, Mutchler, and Hopwood (1991a) identified 134 NYSE and ASE companies that filed for bankruptcy between 1974 and 1985 and found that 54 of the companies (40.3%) in their sample received a going concern modification in the last audit report issued prior to bankruptcy.[2]

Chen and Church (1992) provided the most recent evidence of auditor reporting for bankrupt companies. They identified 53 companies that filed for bankruptcy between 1983 and 1987 and found that 22 of the companies (41.5%) received a going concern modification in the period prior to bankruptcy.[3]

The Impact of SAS 59

The debate within the profession about auditor's reporting for going concern uncertainties did not end with the issuance of SAS 34. Mutchler (1984) interviewed executive-level partners of the

Table 1. Prior Research on the Audit Reports Received by Bankrupt Companies in the Year Prior to Bankruptcy

	Altman and McGough (1974)	Altman (1982)	Menon and Schwartz (1986)	McKeown et al. (1991a)	Chen and Church (1992)
Bankrupt companies	28	81	147	134	53
Companies that received modified audit reports	13	39	63	54	22
Companies that received unmodified audit reports	15	42	84	80	31
Proportion of bankrupt companies that received a modified report	46.4%	48.1%	42.9%	40.3%	41.5%
Sample selection criteria	Not disclosed	Not disclosed	ASE and NYSE companies listed as bankrupt in the WSJ Index	ASE and NYSE companies listed as bankrupt in the WSJ Index	ASE and NYSE companies listed as bankrupt in the WSJ Index
Years covered	1970-1973	1974-1982	1974-1983	1974-1985	1983-1987

(then) eight largest accounting firms and documented a wide diversity of opinions. While some of the partners indicated that the auditor, by assessing the going concern status of a company, was in a position to provide a signal to users, others flatly rejected this viewpoint. There was also disagreement over the role that the recoverability of assets should play in the reporting decision. Several partners indicated that the auditor's responsibility was to provide a signal to users about potential going concern problems regardless of the recoverability of assets. Others indicated that the recoverability concept was theoretically correct, but that it was too difficult to measure to be useful. Finally, some partners felt that the recoverability concept was both theoretically correct and useful in practice.

While the debate within the profession continued, a number of high profile business failures, some following unqualified audit reports, raised a public cry of "where were the auditors" (Connor 1986). Justly or unjustly, the public clearly viewed these business failures as audit failures (Berton 1985; Connor 1986). These business failures also attracted the attention of Congress and the financial press. In remarks before Congress, Congressman Wyden stated: "In one financial disaster after another... the disaster struck virtually on the heels of clean audit certificates issued by audit firms indicating that the companies were financially sound." The erosion of the public's confidence in the audit profession in the mid-1980s led one major accounting firm to suggest that auditors address the public's concerns by taking on the responsibility of assessing a company's financial condition.

In this environment of eroding public confidence, congressional scrutiny, and debate within the profession, the ASB began deliberations on a set of audit standards, including one that addressed the auditor's responsibility to evaluate and report on a client's ability to continue as a going concern. Kaplan and Pany (1992) reviewed the comment letters to the exposure draft and found that there was substantial disagreement about the proposed standard. The (then) eight largest accounting firms were split almost evenly, and a similar split was observed among other respondents. In the final vote, three of the (then) Big Eight voted against the standard.

SAS 59, issued in April of 1988 and effective for the audits of fiscal years beginning on or after January 1, 1989, increased the auditor's responsibilities in two ways. First, under *SAS 34* the entity's continuation was assumed and the auditor was required to consider the going concern issue only when normal audit procedures produced "contrary information." Ellingsen, Pany, and Fagan (1989) refer to this as a negative duty. Under *SAS 59*, the auditor is required to evaluate the entity's ability to continue as a going concern on every engagement, not just engagements where contrary information is uncovered. Thus, *SAS 59* increased the auditor's responsibility by imposing an affirmative duty.

Second, in addition to increasing the auditor's responsibility to evaluate going concern uncertainties, *SAS 59* increased the auditor's responsibility for reporting such uncertainties. Under *SAS 34*, an auditor was required to modify the audit report only if (1) there was substantial doubt about an entity's ability to continue as a going concern and (2) the recoverability of asset values was questionable. Thus, a company with financial difficulties would not necessarily receive a going concern opinion if the auditor believed that the recorded asset values were recoverable. Alternatively, under *SAS 59* the auditor must modify the audit report whenever substantial doubt exists about the entity's ability to continue as a going concern, regardless of the recoverability of asset values.

While *SAS 59* altered the conditions that would lead the auditor to issue a going concern audit report, *SAS 58*, "Reports on Audited Financial Statements" (AICPA 1988a) simultaneously altered the form of the report. Previously under *SAS 2*, the auditor issued a "subject to" opinion qualification for a material uncertainty, including uncertainty regarding an entity's going concern status. *SAS 58* eliminated the "subject to" opinion qualification and required that material uncertainties, including going concern uncertainties, be reported in a separate explanatory paragraph after the opinion.

Testable Hypothesis

The testable hypothesis in this study is developed within the evaluation research paradigm. The purpose of evaluation research is to assess the descriptive validity of claims made about a phenomenon (Simon and Burstein 1985). An example of evaluation research would be a study that assesses the claims made about school busing. Prior to busing, advocates made claims that busing would improve students' educational attainment and socialization. A study that evaluates whether such improvements actually come to pass would be considered evaluation research.

Evaluation research is conducted frequently in social sciences; however Frost and Kinney (1993) recently used the evaluation research paradigm in accounting to investigate whether the SEC, through changes in the registration process, achieved its objective of increasing the comparability of information about foreign registrants. Similarly, this research attempts to assess whether, subsequent to *SAS 59*, more bankrupt companies received modified audit reports prior to the bankruptcy. Connor (1986) noted that this issue is at the heart of the expectation gap.

From the studies summarized in Table 1 that previously examined the relationship between bankruptcy and going concern audit reports issued prior to *SAS 59*, less than one-half of the bankrupt companies received a modified audit report in the period prior to bankruptcy. *SAS 59*, however, increased the auditor's responsibilities for evaluating and reporting on a client's ability to continue as a going concern. Consequently, some companies that received unmodified audit reports under *SAS 34* (either because the normal audit procedures did not produce contrary information or because the recorded asset values were judged to be recoverable) may receive going concern modifications under *SAS 59*. Thus, the proportion of bankrupt companies receiving a modified audit report in the period prior to bankruptcy should increase after *SAS 59*.[4] This hypothesis is stated below (in the alternative form).

Hypothesis 1.   The proportion of bankrupt companies receiving a modified audit report in the year prior to bankruptcy will be larger in the time period covered by *SAS 59* than in the time period covered by *SAS 34*.

RESEARCH APPROACH

We use a cohort design described in Cook and Campbell (1979) to evaluate the impact of *SAS 59*. In the context of our study, the two cohorts are the sample of bankrupt companies whose audit reports were issued before the effective date of *SAS 59* and the sample of bankrupt companies whose audit reports were issued after the effective date of *SAS 59*.

Sample Selection

The initial sample of bankrupt companies consisted of 435 companies that filed for bankruptcy between December 1986 and March 1992. The sample was identified from the *Wall Street Journal Index* and various regional and national wire services from NEXIS. Companies were then deleted from the initial sample if either (1) their financial statements could not be located through sources available to the authors (190 companies), or (2) their last audit report prior to the bankruptcy was issued during the transition period when the auditor had the option of following either *SAS 34* or *SAS 59* (60 companies).[5] These deletions resulted in a final sample of 185 bankrupt companies. The date of bankruptcy filing for each of the 185 companies in the final sample was determined and the most recent audit report dated before the bankruptcy filing date was examined to determine the type of report that each company received and to ensure that the actual filing had not occurred prior to the end of fieldwork.

Classification of Audit Reports

An audit report was classified as modified if, due to expressed doubts about the entity's ability to continue as a going concern, the auditor: (1) disclaimed an opinion, (2) qualified the audit report (only applicable in the *SAS 34* period), or (3) added an explanatory paragraph to the report (only applicable in the *SAS 59* period). Although the wording of the explanatory paragraphs after *SAS 59* were fairly consistent, the phrasing of the paragraphs accompanying the "subject to" qualifications was diverse. Following McKeown et al. (1991a), reports from the *SAS 34* period were considered modified if the auditor expressed doubts about the entity's ability to finance future operations or about the recoverability and classification of assets.

ANALYSIS AND DISCUSSION OF RESULTS

Table 2 presents summary descriptive information about the incidence of modified reports before and after *SAS 59* for the bankrupt companies. The two rows (*SAS 34* and *SAS 59*)

Table 2. Summary of Association Between Audit Opinion
Type and Period: Bankrupt Companies

| | Audit Opinion Type | | |
Period	Modified	Unmodified	Total
SAS 34	36	42	78
	(46.1%)	(53.9%)	
SAS 59	61	46	107
	(57.0%)	(43.0%)	
Total	97	88	185
	$Z = 1.46$ p-value < 0.08		

denote whether the audit report was issued in the period when *SAS 34* or *SAS 59* was the authoritative standard. The observed proportion of bankrupt companies receiving modified audit reports in the *SAS 34* period was 46.1%. Consistent with the results of prior research, less than one-half of the bankrupt companies in the *SAS 34* period received a going concern modification in the year prior to the bankruptcy filing. In the *SAS 59* period, however, the proportion of bankrupt companies receiving a modified audit report prior to the bankruptcy filing increased to 57%. A Fisher-exact test was used to determine whether the proportion of modified audit reports for bankrupt companies was significantly larger in the *SAS 59* period than in the *SAS 34* period. The test indicates that, consistent with Hypothesis 1, the observed increase is statistically significant ($p < 0.08$).

As discussed previously, the design used to investigate the impact of *SAS 59* is a cohort design. Illusory correlation is of particular concern with a cohort design because the "quasi-comparability" of cohorts is assumed. However, a cohort design does not rule out selection differences the way that random assignment does. Accordingly, the design is strengthened if third variables that may be causally linked to the dependent variable can be identified and reliably measured.

The results of prior research suggest that two important third variables, financial condition and size, may differ between the two cohorts. Several prior studies have found that financial condition is the most important explanatory factor in determining whether a company receives a modified audit report. Similarly, prior studies (e.g., Mutchler 1986; McKeown et al. 1991a; Chen and Church 1992) have documented a client-size effect on the auditor's decision to issue a going concern opinion. Generally, this research has found that smaller companies and companies in poorer financial condition are more likely to receive a going concern modification to their audit reports.

The threat to our ability to attribute any observed differences (e.g., in the proportion of modified audit reports received by bankrupt companies) to *SAS 59* stems from the possibility that the sample bankrupt companies in the *SAS 59* period may differ from the sample bankrupt companies in the *SAS 34* period with respect to size and/or financial condition. If the bankrupt companies in the *SAS 59* period either are smaller or are in poorer financial condition than the bankrupt companies in the *SAS 34* period, an increase in modified audit reports might be observed that would not be attributable to *SAS 59*.

To reduce the likelihood that the results reported in Table 2 are attributed to illusory correlation, we used a logistic regression model in which financial condition and size are measured and used as control variables.[6] The explanatory power (pseudo-R^2) of the logistic model was 0.16 and the overall model chi-square was significant at the .01 level. Thus it appears that the model was useful in explaining the issuance of going concern modifications (which is consistent with prior research). Analysis of the logistic results provide strong support for Hypothesis 1. Results indicate that after controlling for financial condition and company size, auditors are more likely to issue a modified audit report for bankrupt companies in the *SAS 59* period than in the *SAS 34* period.

DISCUSSION

Prior to discussing the results of this study, certain data and statistical limitations should be noted. First, the financial condition score used as a proxy for financial risk is imperfect, and therefore it may add noise to our statistical analysis. Second, while we have attempted to control for differences over time in the size and financial condition of sample companies, changes in other unspecified variables may have occurred. We cannot rule out the possibility that a change in an unspecified variable may partially account for the observed results.

Subject to the limitations discussed above, the results of this study support the hypothesis that the proportion of bankrupt companies receiving modified audit reports in the period prior to bankruptcy increased subsequent to *SAS 59*. A logistic regression that controlled for size and financial condition yielded similar results, suggesting that the observed increase was not due to differences in size or financial condition.

A question may be raised regarding why the increase in modified reports noted for bankrupt companies is not larger. Specifically, there are still a substantial number of cases where bankruptcies are preceded by unmodified audit reports. McKeown et al. (1991a) define a Type II error as an unmodified opinion given to a firm that subsequently files for bankruptcy (prior to the next years' audit). A fundamental criticism of this error definition is that professional standards do not equate the auditor's going concern decision with the prediction of bankruptcy.

McKeown et al. (1991a) suggest that firms that have not exhibited signs of financial stress (and have previously received unqualified audit reports) may be pushed into bankruptcy by a sudden unforeseen event. An auditor in such a case who did not have substantial doubt about the company's ability to continue as a going concern (due to the lack of financial stress indicators) would issue an unmodified opinion. This case would be classified as a Type II error according to McKeown et al.'s definition although the auditor has acted in a manner consistent with professional standards.

Accordingly, while the audit reports received by bankrupt companies is a useful measure of auditor reporting behavior, caution must be exercised in post-hoc classifications of audit reports as "errors." However, McKeown et al. (1991b) contend that these cases are errors from a user's perspective, and are at the heart of the expectation gap. Regardless of whether such instances are defined as errors, they clearly can impose costs on society including the auditor. For example, an auditor that issues an unmodified opinion prior to a bankruptcy may suffer from loss of reputation and litigation relating to losses suffered by investors and creditors.

Some conjectures can be offered for why auditor performance in this area is not better. First, it is important to note that some bankruptcies may not be predictable. Argenti (1976) notes that some companies fall into bankruptcy suddenly, with no apparent prior signs of financial distress. The auditor is probably least likely to issue a going concern modification for these companies since there was no prior evidence of financial distress.

Second, the auditor has limited relative expertise in the judgments necessary under *SAS 59*. While it is reasonable to assume that auditors should recognize financial difficulties, *SAS 59* does not suggest that all companies with financial difficulties be issued modified audit reports. In fact, auditors must consider mitigating factors and management's plan for addressing the financial difficulties. According to *SAS 59*, the auditor's goal in reviewing management's plan is twofold. First, the auditor must assess whether the plan (assuming it is implemented) will effectively mitigate the conditions or events that resulted in substantial doubts about the company's going concern status. Second, the auditor must evaluate whether it is likely that management's plan can or will be implemented.

The auditor may not have sufficient experience or training in either of these tasks (assessing the efficiency of operating plans or determining the likelihood of future events) to develop any degree of expertise. Additionally, Carmichael and Pany (1993) note that there is little authoritative guidance available to direct auditors in reviewing management's plan.

SAS 59 leaves the auditor in a difficult position. If, based on the evidence collected, the auditor has substantial doubts about a company's going concern status, management is informed of the auditor's doubts and is required to prepare a plan. Management, who presumably does not want the company's financial stability questioned in the auditor's report, then has the opportunity to produce a report that they believe will allay the auditor's doubts. Given the auditor's limited expertise it may be difficult to argue convincingly that the plan is insufficient to deal with the company's financial difficulties.

CONCLUSIONS

Prior research has documented that less than 50% of bankrupt companies receive a going concern audit report prior to bankruptcy. Carmichael and Pany (1993) suggest that one of the main elements of the public's dissatisfaction with the audit profession was the public's expectation that the auditor will warn them about impending business failure.

As one of the expectation gap auditing standards, the intent of *SAS 59* was to address the public's doubt by increasing the auditor's responsibility regarding the going concern status of clients. Carmichael and Pany (1993) called for research to determine whether the auditor's performance (in giving going concern opinions to bankrupt companies) has improved subsequent to the passage of *SAS 59*. The results of this study provide evidence that the proportion of failed companies receiving modified audit reports in the period prior to bankruptcy increased subsequent to *SAS 59*.

APPENDIX A

Appendix A presents details of the logistic regression used to test our hypothesis. As was discussed above, our ability to draw inferences regarding the impact of *SAS 59* is strengthened by measuring and controlling for the size and financial condition of bankrupt companies. SIZE was measured as the natural log of company sales. The measure of financial condition (FC) used in this study is based on the multivariate failure prediction model developed by McKeown et al. (1991a). This model measures the probability of bankruptcy as a function of a set of seven predictive financial variables such as current ratio, long-term debt to total assets ratio, and so on. Higher scores for the probability of bankruptcy reflect greater financial distress. Conversely, lower scores indicate a company with lower financial risk. Because Mckeown et al.'s parameter estimates were derived based on companies from the 1974 to 1985 period, we re-estimated the model parameters after retrieving the necessary financial data from Compustat for the sample of bankrupt companies and a random sample of 400 companies that did not declare bankruptcy between December 1986 and March 1992.[7] Because some of the predictive financial variables such as the current ratio could not be computed for 59 bankrupt companies, the multivariate failure prediction model was estimated with a reduced sample of 126 bankrupt companies.[8]

Both size and financial condition were used as control variables in the following logistic model (a statistical technique that is appropriate in instances when the operational measure of the dependent variable is dichotomous):

Table A1.

Variable	Description of the Variable
REPORT	A dummy variable indicating whether a firm received a modified opinion,
SIZE	Natural log of sales,
FC	A continuous measure of financial condition using a multivariate failure prediction model,
POST	Coded "1" if SAS 59 was in effect at the time the audit report was issued, "0" otherwise,
FC__POST	Represents the interaction of FC and POST,
SIZE__POST	Represents the interaction of SIZE and POST,

$$REPORT = \beta_0 + \beta_1*SIZE + \beta_2*FC + \beta_3*SIZE\_POST + \beta_4*FC\_POST +$$

$$\beta_5*POST + e$$

where all variables are defined in Table A1, $\beta_1 - \beta_5$ represent coefficient associated with the variables, and e represents the error term.

Both control variables (SIZE and FC) were allowed to interact with the POST variable, which takes on the value "1" if SAS 59 was in effect at the time the audit report was issued, and "0" otherwise. The introduction of these two interaction terms (SIZE__ POST and FC__POST) into the model allows the relationship between the dependent variable (REPORT) and the control variables (SIZE and FC) to change between the SAS 34 and SAS 59 periods. Ignoring these interactions would have imposed an artificial constraint.

POST, the test variable used to assess whether the incidence of modified audit reports increased in the SAS 59 period, is coded as "1" if SAS 59 was in effect at the time the audit report was issued, and "0" otherwise. Given the control variables in the model, the coefficient β_5 relates to bankrupt companies whose audit reports were issued when SAS 59 was in effect. A significant positive (negative) coefficient indicates that bankrupt companies were more (less) likely to receive a modified audit report in the SAS 59 period. Because Hypothesis 1 predicts that more bankrupt companies will receive modified reports after SAS 59, the statistical test is whether β_5 is significantly greater than zero.

Table A2. Logistic Regression Results
(N = 126)

Variables	Parameter Estimate	Chi-square Statistic
Intercept	3.67	5.48**
SIZE	0.15	0.75
FC	6.21	9.05***
SIZE_POST	-0.33	1.88
FC_POST	-2.70	1.21
POST	3.16	2.71**
Pseudo-R^{2a}	0.16	
Likelihood ratio statistic[b]	23.31***	

Notes: All variables are as defined in Table A1.
$** p < 0.05$
$*** p < 0.01$
[a] The pseudo-R^2 equals 1 – (log likelihood at convergence/log-liklihood at zero).
[b] The likelihood ratio statistic is computed to test the hypothesis that all the parameters in the model are simultaneously equal to zero. Under this null hypothesis, the statistic has an asymptotic distribution which is a chi-square with the degrees of freedom equalling the number of parameters in the model.

Logistic Results

The logistic results reported in Table A2 are based on 126 bankrupt companies.[9] The explanatory power of the logistic model (pseudo-R^2) is 0.16. The log likelihood ratio statistic of 23.3, as reported in the bottom of Table A2, indicates that the hypothesis that all logistic coefficients are simultaneously equal to zero is rejected at the 0.01 significance level.

The financial condition variable is significant with the expected positive sign ($p < 0.01$). Consistent with the findings of prior research, companies in poorer financial condition are more likely to receive a going concern modification. Alternatively, the client size variable is not significant. Neither of the interaction terms is significant suggesting that the relationship between SIZE and financial condition variables and the likelihood of receiving a going concern modification did not change significantly between the *SAS 34* and *SAS 59* periods.

As hypothesized, and consistent with the evidence presented in Table 2, β_5, the coefficient on the POST variable is 3.16 and is significant at the .05 level. The coefficient indicates that after controlling for financial condition and company size, auditors are more likely to issue a modified opinion for bankrupt companies in the *SAS 59* period than in the *SAS 34* period. Thus, the test of β_5 supports Hypothesis 1.

ACKNOWLEDGMENT

The authors wish to acknowledge the helpful comments of Linda Johnson, Steve Kaplan, Kurt Pany, Kurt Reding, and Earl Wilson on earlier drafts of this manuscript.

NOTES

1. *SAS 59* was issued in April of 1988 and became effective for the audits of fiscal years beginning on or after January 1, 1989.
2. McKeown et al. (1991a) also classified the sample of bankrupt firms by signs of financial stress. Of the 134 bankrupt firms, 16 were classified as nonstressed. None of the 16 nonstressed firms received a modified audit report prior to bankruptcy.
3. Chen and Church (1992) also considered the default status of the bankrupt companies at the time of the audit report. Twenty-three of the 53 bankrupt companies were classified as in default at the time of the audit report. Twenty-one of these companies received a modified audit report. Alternatively, of the 30 bankrupt companies that were not in default as of the audit report date, only one received a modified audit report.
4. Although *SAS 59* is referred to throughout the remainder of the paper, both *SAS 58* and *SAS 59* became effective simultaneously. Because *SAS 58* altered the form rather than the substance of the auditor communication, tests assessing the joint impact of *SAS 58* and SAS 59 should not be problematic.
5. *SAS 59* was released in April, 1988 and was effective for audits of fiscal years beginning on or after January 1, 1989 (with earlier application permitted). The audit reports of 60 sample companies were issued after April 1988, but covered financial statements of a period ending before December 31, 1989. The auditor's responsibility to assess a client's going concern status during this transition period depended on whether the auditor followed *SAS 34* or *SAS 59*. Because it is impossible to identify which standard a given auditor followed during the transition period, the companies in the transition period were dropped from the analysis.
6. For a detailed discussion of the logistic model, please see Appendix A.
7. The final model used to predict client financial condition is given as:

0.359 + 0.2357*(NET INCOME/TOTAL ASSETS) - 0.4524*(CURRENT ASSETS/SALES) - 1.701*(CURRENT ASSETS/CURRENT LIABILI-

TIES) + 3.575*(CURRENT ASSETS/TOTAL ASSETS) - 0.738*(CASH/
TOTAL ASSETS) + 2.644*(LONG-TERM DEBT/TOTAL ASSETS) -
0.171*(NATURAL LOG OF SALES).

The explanatory power of the logistic model is 0.25 which is comparable to the
estimates provided by McKeown et al. (1991a).

8. As an example, some companies (such as banks) did not report classified
balance sheets. Accordingly, current ratio could not be computed.

9. To assess potential biases, if any, arising from reduction in the sample of
bankrupt firms, we examined the incidence of modified audit reports and industry
concentration for the reduced sample over the two time periods. For the reduced
sample of 126 bankrupt companies, 24 of the 52 bankrupt companies (46.1%)
received a modified audit report in the *SAS 34* period. In the *SAS 59* period,
however, 41 of the 74 bankrupt companies (55.4%) received a modified audit report
prior to the bankruptcy filing. These proportions are similar to those reported in
Table 2 for the full sample, thereby reducing the possibility of either understating
the proportion of modified audit opinions in the *SAS 34* period or overstating the
proportion of modified audit opinions in the *SAS 59* period for the reduced sample.
Similarly, classification of the sample by Standard Industrial Classification (SIC)
codes indicated a wide cross-section of industries and no particular industry grouping
was dominant.

REFERENCES

Altman, E. 1982. Accounting implications of failure prediction models. *Journal of
Accounting, Auditing, and Finance* (Fall): 4-19.

Altman, E., and T. McGough. 1974. Evaluation of a company as a going concern.
Journal of Accountancy (December): 51-7.

American Institute of Certified Public Accountants. 1942. *Statement on Auditing
Procedure No. 15*, Disclosure of the Effect of Wartime Uncertainties on
Financial Statements. New York: AICPA.

_____. 1963. *Statement on Auditing Procedure No. 33*, Auditing Standards and
Procedures: A Codification. New York: AICPA.

_____. 1974. *Statement on Auditing Standards No. 2*, Reports on Audited
Financial Statements. New York: AICPA.

_____. 1981. *Statement on Auditing Standards No. 34*, The Auditor's
Considerations When a Question Arises about an Entity's Continued
Existence. New York: AICPA.

_____. 1988a. *Statement on Auditing Standards No. 58*, Reports on Audited
Financial Statements. New York: AICPA.

_____. 1988b. *Statement on Auditing Standards No. 59*, The Auditor's
Consideration of an Entity's Ability To Continue as a Going Concern. New
York: AICPA.

Argenti, J. 1976. *Corporate Collapse: The Cause and Symptoms.* New York: John
Wiley & Sons.

Berton, L. 1985. Accountants aim to prevent audit failures. *The Wall Street Journal,* February 19, p. 4.

Carmichael, D., and K. Pany. 1993. Reporting on uncertainties, including going concern. In *The Expectation Gap Standards: Progress, Implementation Issues, Research Opportunities,* 35-55. New York: AICPA.

Chen, K., and B. Church. 1992. Default on debt obligations and the issuance of going-concern opinions. *Auditing: A Journal of Practice & Theory* (Fall): 30-49.

Commission on Auditors' Responsibilities. 1978. *Report, Conclusions, and Recommendations.* New York: AICPA.

Connor, J. 1986. Enhancing public confidence in the accounting profession. *Journal of Accountancy* (July): 76-85.

Cook, T., and D. Campbell. 1979. *Quasi-Experimentation Design and Analysis Issues for Field Settings.* Boston, MA: Houghton-Mifflin.

Ellingsen, J., K. Pany, and P. Fagan. 1989. SAS No. 59: How to evaluate going-concern. *Journal of Accountancy* (January): 24-31.

Frost, C., and W. Kinney. 1993. Regulation S-X and comparability of disclosure for foreign registrants in the U.S. Working paper, Washington University, St. Louis, MO.

Guy, D., and J. Sullivan. 1988. The expectation gap auditing standards. *Journal of Accountancy* (April): 36-46.

Holstrum, G. 1991. ASB moves forward on projects—seeks research input on expectation gap issues. *The Auditor's Report* (Fall): 16-8.

Kaplan, S., and K. Pany. 1992. A study of public comment letters on the auditor's consideration of the going-concern issue. In *Research in Accounting Regulation,* Vol. 6, ed. G.J. Previts, 3-23. Greenwich, CT: JAI Press.

McKeown, J., J. Mutchler, and W. Hopwood. 1991a. Towards an explanation of auditor failure to modify the audit opinions of bankrupt companies. *Auditing: A Journal of Practice & Theory* (Supplement): 1-20.

_____. 1991b. Reply. *Auditing: A Journal of Practice & Theory* (Supplement): 21-24.

Menon, K., and K. Schwartz. 1986. The auditor's report for companies facing bankruptcy. *The Journal of Commercial Lending* (January): 42-52.

Mutchler, J. 1984. Auditor's perceptions of the going-concern opinion decision. *Auditing: A Journal of Practice & Theory* (Spring): 17-30.

_____. 1986. Empirical evidence regarding the author's going-concern opinion decision. *Auditing: A Journal of Practice & Theory* (Fall): 148-163.

Securities and Exchange Commission. 1962. *Accounting Series Release No. 90, Certification of Financial Statements.* Washington, DC: SEC.

Simon, J., and P. Burstein. 1985. *Basic Research Methods in Social Science,* 3rd ed. New York: Random House.

THE EUROPEAN UNION:

REGULATION MOVES FINANCIAL REPORTING TOWARD COMPARABILITY

Kathleen R. Bindon and Helen Gernon

ABSTRACT

The main objective of the European Union (EU) is the creation and development of a Common Market through the free flow of goods, persons and capital. Within this context, one of the goals was to harmonize financial reporting practices across the 12 member states in order to make these practices more comparable. This move toward harmonization was accomplished through the issuance of Directives that were legally binding. Thus, the EU was in the unique position of being able to use a legal framework to harmonize the financial reporting practices of a particular region of the world. However, the designers of the Directives took a mutual recognition approach to developing their content. This approach allows for more diversity and results in less comparability. Ultimately, mutual recognition of cultural differences, accounting values, and regulatory environments

Research in Accounting Regulation, Volume 9, pages 23-48.
Copyright © 1995 by JAI Press Inc.
All rights of reproduction in any form reserved.
ISBN: 1-55938-883-8

across the 12 EU countries was necessary to achieve adoption of the Directives by the member countries. This paper addresses the question of why the EU took a mutual recognition approach given that it was in the unique position of having the opportunity to use a legal framework to enhance the comparability of financial reporting practices. The paper explores and explains why achieving comparability of financial reporting practices could not be expected to result from the Directives. The paper helps the reader understand that comparability could not be achieved at this time due to the persistent underlying differences in cultural attitudes, accounting values and individual regulatory environments across the 12 member states. The harmonization argument is developed by applying, extending, and integrating prior classification models to explain the persistent diversity in accounting measurement, valuation, and disclosure practices that exist across the EU countries. Patterns in measurement, valuation, and disclosure have been studied by many researchers in a variety of ways. Other researchers have used more of a Farmer-Richman (1966) approach where environmental analysis is used to explain and understand differences in accounting principles and modes of regulation. This paper integrates selected pieces from the pattern research literature with selected pieces from the environmental analysis research literature to offer an innovative interpretation as to why financial reporting diversity continues to exist in the European Union.

INTRODUCTION

The main objective of the European Union (EU) is the creation and development of a Common Market through the free flow of goods, persons, and capital. Within this context, one of the goals was to harmonize financial reporting practices across the 12[1] member states (Belgium, Denmark, France, Germany, Greece, Ireland, Italy, Luxembourg, the Netherlands, Portugal, Spain, and the United Kingdom). This move toward harmonization was accomplished through the issuance of Directives that were legally binding on these member states. Thus, the EU was in the unique position to use a legal framework to harmonize the financial reporting practices of a particular region of the world. Instead, it took a mutual recognition approach. Mutual recognition is "a process by which the regulations in one country are accepted as equivalent in another, subject to the minimum standards set in the 4th and 7th Directives" (Radebaugh and Gray 1993, 159).

This paper addresses the question of why the EU took a mutual recognition approach given that it was in the unique position of having the opportunity to use a legal framework to enhance the comparability of financial reporting practices. The paper explores and explains why achieving comparability of financial reporting practices could not be expected to result from the Directives of the EU regulatory agencies. The paper helps the reader understand that comparability could not be achieved at this time due to the persistent underlying differences in cultural attitudes, accounting values and individual regulatory environments across the 12 member states.

Much has been written about accounting harmonization and standardization (Mueller 1991; Purvis, Gernon, and Diamond 1991; Van Hulle 1989a,b). It is important to distinguish between these two terms, which are not interchangeable, as well as to distinguish between the related concepts of harmony and uniformity. As defined by Tay and Parker (1990, 72):

> Harmonization (a process) is a movement away from total diversity of practice. Harmony (a state) is therefore indicated by a 'clustering' of companies around one or a few of the available methods. Standardization (a process) is a movement towards uniformity (a state). It includes the clustering associated with harmony, and reduction in the number of available methods.

As can be seen from these definitions, harmony and uniformity are not dichotomous states. Total diversity and uniformity are the extreme states, at opposite ends of the continuum, whereas harmony, as defined by Tay and Parker, is any point along the continuum, excluding the extreme states of total diversity and uniformity. This implies that the state existing after even a small movement away from total diversity would be harmony. This definition of harmony seems to be too broad to use for classification and discussion purposes.

As an alternative to Tay and Parker's continuum, Exhibit 1 offers a subjective division of harmony into two states, bounded diversity and comparability. The state of bounded diversity exists at any point along the continuum from the extreme state of total diversity to the continuum's midpoint. The state of comparability, then, begins at the midpoint and continues until reaching the extreme state of total uniformity. It is into the state of bounded diversity,

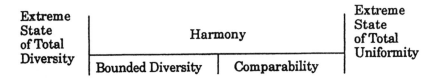

Source: Adapted from Tay and Parker (1990).

Exhibit 1.   Continuum of Total Diversity to Total Uniformity

not comparability, that the EU, through the issuance of Directives, has been able to move the accounting and reporting practices of its member countries.

The 4th and 7th Directives had the most impact on financial reporting practices. The 4th addressed valuation principles and disclosure practices, while the 7th dealt with the accounting issues of consolidation. The designers of the Directives took a mutual recognition approach to developing the contents of the Directives. Ultimately, mutual recognition of cultural differences, accounting values, and regulatory environments across the 12 EU countries was necessary to achieve adoption of the Directives by the member countries. This mutual recognition approach allowed the writers to design Directives that were politically palatable, addressing broad issues and setting minimum requirements rather than providing detailed guidance. By choosing not to provide more detailed guidance, which would have reduced allowable options, the Directives could not be expected to achieve the goal of comparability of financial reporting practices across the EU member countries but were able to move from total diversity to bounded diversity.

The harmonization argument is developed by applying, extending, and integrating prior classification models to explain the persistent diversity in accounting measurement, valuation, and disclosure practices that exist across the EU countries. Patterns in measurement, valuation, and disclosure have been studied by many researchers in a variety of ways (Nair and Frank 1980; Nobes 1983; Gray 1988). Other researchers have used more of a Farmer-Richman (1966) approach where environmental analysis is used to explain and understand differences in accounting principles and modes of regulation (Mueller 1967; Hofstede 1987; Puxty, Willmott, Cooper, and Lowe 1987). Kirsch (1994) incorporated both pattern classification and environmental analysis in examining the influence of cultural differences on securities regulation and on accounting reporting and disclosure standards. In his work, Kirsch looked at two major cultural blocks, the Anglo-American and the Chinese-Asian. To date, no research of this nature has been undertaken which focuses solely on EU countries. As the EU was in a unique position to utilize its power to force accounting comparability on member countries with very diverse cultures, it is important to investigate what caused the EU to elect not to force this comparability among its member states. These causes need to be explored and understood as they may have significant importance to any worldwide harmonization efforts. Therefore, this paper integrates selected pieces from the pattern research literature with selected pieces from the environmental analysis research literature to offer an innovative interpretation as to why financial reporting diversity continues to exist among the EU countries.

The paper is organized as follows. The first section provides a summary of the 4th and 7th Directives. A review of existing research which addressed variables that shape accounting practice is presented in the second. The third section analyzes and extends selected prior classification studies. The fourth section applies the knowledge gained by integrating the prior classification studies to understanding why the differences in valuation, measurement, and disclosure practices exist across the EU countries. The fifth section presents a summary and conclusions, as well as ideas for future research.

BOUNDED DIVERSITY AND THE 4TH AND 7TH DIRECTIVES

The EU attempted to use a legal framework to achieve a minimum and acceptable level of comparability across financial statements compiled in each of the 12 different member states. One of the reasons given by the Commission for reducing the number of existing financial reporting differences was to enhance the efficient movement of capital throughout the EU. Several researchers are currently investigating whether comparability does impact the efficient movement of capital (Biddle and Saudagaran 1989; Choi and Levich 1990).

This paper suggests that there is a difference between comparability and bounded diversity as the latter state allows for more diversity than the former. When the financial reporting Directives were being drafted, the Commission was in the unique position of being able to legally mandate a move toward financial reporting comparability (a state past the midpoint on the continuum from total diversity to total uniformity). However, a closer look at the 4th and 7th Directives allows us to see that they fell short of comparability. However, they do achieve the worthy goal of bounded diversity by reducing the extent of national financial reporting differences across member states.

This paper explores and explains why the authors of the EU Directives took a compromise position while drafting the 4th and 7th Directives, resulting in bounded diversity rather than comparability. National financial accounting and reporting practices closely follow national company law. The harmonization of EU company law was an attempt to reduce the diversity in existing national differences in financial accounting and reporting. While the Directives could not reduce diversity sufficiently to move financial accounting and reporting into a state of comparability, they were able to achieve a minimum degree of harmony, bounded diversity.

The move toward bounded diversity was accomplished through the issuance of Directives. Directives were authorized by the Treaty of Rome as a way to harmonize national company laws. Directives were issued by the European Commission and if adopted by the EU Council of Ministers, were binding on the member states. It was mandatory that each Directive be incorporated into member states' national laws through implementing legislation. However, each country had several degrees of freedom to choose from when selecting the form and method of implementation. In addition, the Directives were intended to be a minimum rule which could be strengthened at the national level. In certain cases, the Directives allowed a country to choose between two or more alternative financial reporting options.

When countries that have diverse cultural attitudes, diverse accounting values, and diverse regulatory environments are given choices in what to measure, how to value, and what to disclose, financial reporting practices will remain diverse. Due to the available options countries could select among in their implementing

Exhibit 2. Valuation Principles and
Reporting Practices Adopted by the EU 4th Directive

Valuation Principles

The 4th Directive adopts several general principles of valuation that must be adhered to in the preparation of financial statements. Among them are the following:

- the company must be presumed to be a going concern;

- methods of valuation must be consistent from year to year;

- profit calculations must be based on the concept of prudence (including any profits earned as of the balance sheet date while also including all foreseeable losses); and

- income and expenses must conform to the matching principle and be calculated on the accrual basis.

Other Reporting Practices

The 4th Directive sets out a number of other minimum standards for reporting covering a variety of areas. Following are some of the highlights.

- Companies generally must use historic cost as the basis for asset valuation, but member states may allow replacement value accounting for certain fixed assets so long as it requires full disclosure of the impact on the financial statements.

- Companies need not disclose the effects of inflation on financial statements unless required by the member state. Any revaluation reserve created by the application of methods of accounting for inflation may not be distributed as dividends unless realized.

- If a member nation allows the capitalization of research and development expenses, any capitalized amount should be written off in five years or less. Longer periods may be allowed under exceptional circumstances if appropriate disclosure is made in the notes.

- Taxes on ordinary profit and extraordinary items should be disclosed separately, though the two may be combined on the face of the profit-and-loss statement. When income tax expense differs materially from income tax payable, the difference must be disclosed. Deferred tax accounting is not required.

- Pension costs charged against income must be disclosed. However, the Directive contains no detailed guidance for determining pension expenses or liabilities.

- Subscribed capital, paid-in capital, revaluation reserve (if applicable), other reserves, and retained earnings must be disclosed on the balance sheet. The number and par value of shares issued during the year, the number and par value of each class of stock outstanding, and a description of convertible debentures, stock rights, or other such instruments must be disclosed in the notes.

- The 4th Directive makes no specific requirements with regard to accounting for leases, related party transactions, or foreign currency translation. The only segment reporting requirements are net sales broken down by line of business and geographical market, and average number of employees by category.

Source: Haskins (1991).

legislation, the Directives could not achieve comparability but could move financial reporting practices further away from total diversity into the bounded diversity range and closer to comparability.

The 4th Directive governs the form and content of financial statements and provides minimum requirements as to the content of the notes to the financial statements and the annual report. In addition, the 4th Directive adopted the true and fair view criterion for preparation and presentation of financial statements. Exhibit 2 provides a summary of the valuation principles and reporting practices adopted by the 4th Directive.

As Exhibit 2 suggests, the 4th Directive took a mutual recognition approach to establishing financial reporting practices by setting minimum standards which encompassed much of the existing diversity of practice among the 12 member countries. The Directive was a politically palatable compromise that moved the member states into the bounded diversity range, but not past the continuum midpoint of comparability. It was more concerned with broad issues rather than detailed rules or the elimination of choices. The drafters of the Directive had to be sensitive to differences in cultural attitudes, accounting values, and regulatory environments to achieve adoption. This is what is meant by mutually recognizing each other's differences.

The 7th Directive established basic rules for preparing consolidated financial statements and generally followed the Anglo-American approach with the emphasis on legal control. Prior to the issuance of this Directive several EU countries had little or no national law that addressed or required the issuance of consolidated financial statements (e.g., Greece, Luxembourg, Italy, and Portugal). After the issuance of this Directive several countries (e.g., Belgium, France, and Germany) had to change their existing consolidation practices because the scope of the 7th Directive included more companies and old accounting methods were no longer appropriate.

So, the final content of the 7th Directive was the result of much negotiating and compromising among countries that had very diverse approaches to the issue of consolidation. The extreme diversity in practices resulted in Directive provisions that provided a minimum level of bounded diversity and disclosure, but not comparability. However, the movement away from almost total diversity toward bounded diversity, which is expected to result from the 7th Directive, is considered to be a much needed improvement in EU reporting standards (Haskins 1991; Mueller 1991; Van Hulle 1989a,b).

PRIOR CLASSIFICATION RESEARCH

Mueller (1967) pioneered the work in international classification of financial reporting as related to business environments. He proposed that accounting practices are the product of their economic, political, legal, and social environments. Using his informed judgment, he identified four patterns of development and illustrated each with one or two examples. His work suggested that the nature of a country's accounting system evolved over time to fit the particular needs of the users of accounting information in that country. It followed that another country's system could not just be "adopted" because the practices would not be appropriate.

Using information contained in Price Waterhouse's 1973 and 1976 *Survey of Accounting Practices in 38 Countries*, Nair and Frank (1980) divided financial reporting practices into those related to disclosure and those related to measurement. The use of cluster analysis showed that groupings of countries changed depending whether you look at disclosure or measurement practices, particularly in the case of Germany. Some years later it was discovered that the Price Waterhouse data contained certain errors and was not designed for cluster analysis.

Nobes (1983) classified developed Western countries by the measurement and valuation practices of their public companies. His work resulted in the development of a model with a hierarchy that showed different groupings of countries based on various factors. More importantly, the hierarchy allowed us to see how close or far away the groups were to each other.

Because of the pioneering classification work of these authors, more attention has been paid to the influence of the environment, and in particular the cultural influence, on the development of financial reporting systems and accounting values (McKinnon and Harrison 1985; Harrison and McKinnon 1986; Hofstede 1987; Gray 1988; Cooke 1991).

By isolating and then classifying only the 12 EU countries according to cultural dimensions, accounting values, and regulatory environments, we begin to understand why comparability among financial reporting practices was not a viable outcome of the Directives. We can see that reducing the diversity of existing practices (bounded diversity) was a worthy, valuable and possible alternative to the original goal of comparability. This approach resulted in recognizing and maintaining country-specific peculiarities, while attaining a minimum standard of valuation and reporting practices across the 12 EU countries.

This paper is interested in the underlying influence of certain environmental variables on financial reporting practices, not with specific differences in financial reporting practices themselves. The work of Hofstede (1980, 1983a, 1983b, 1984, 1987) is used to explore the significance of cultural differences across the 12 EU countries. Gray's (1988) framework that links culture with the development of accounting systems is then applied to only the EU countries. The Puxty et al. (1987) work on modes of regulation is extended to include all of the EU countries and provides insight as to the influence of legal, political, cultural and professional variables on the fundamental structure of each accounting system.

EXTENSIONS OF SELECTED PRIOR CLASSIFICATION RESEARCH AS APPLIED TO THE EU

Cultural Attitudes

While culture can be and is defined in many different ways, for purposes of this paper culture is defined as "the collective programming of the mind which distinguishes one category of people from another" (Hofstede 1987, 1) In a pioneering work which quantified differences in work-related values as a part of national cultures, Hofstede (1980) collected data from over 116,000 questionnaires administered to IBM employees all around the world. The methodology included surveying the same group of employees twice, with anywhere from two to six years between surveys. For 40 countries, there was sufficient data for analysis.[2]

The survey contained both attitude (how do you like your job) and value (preference for one type of boss over another) questions. Answers to the value questions showed remarkable and stable differences between countries (Hofstede 1983b). To find out whether his results were a phenomenon of the corporation initially used, Hofstede repeated the survey to include managers from many different countries attending a business school course he was teaching. He determined that the managers' answers revealed the same patterns of differences in values between countries.

Hofstede's research led to his identification of four principal cultural dimensions along which dominant value systems could be ordered. These dimensions are: Individualism, Power Distance, Uncertainty Avoidance, and Masculinity.[3] Power Distance and Uncertainty Avoidance are most relevant for the functioning of organizations within a country because "the two basic problems of organizing are distributing power and avoiding unwanted uncertainties" (Hofstede 1987, 5). On the other hand, Individualism and Masculinity are more relevant to describing the behavior of individuals within organizations.

Hofstede assigned scores, based on multivariate statistics (factor analysis) and theoretical reasoning, along these four dimensions to each of the countries in his study. Once these scores were determined, Hofstede used cluster analysis to group the countries into culture areas. Hofstede (1987) actually was attempting to quantify national cultures and said of his own work that it represented an accounting approach to culture.

Exhibit 3 is a duplication of Hofstede's plot of the Power Distance and Uncertainty Avoidance scores from his research. However, only the 12 countries of the European Union have been plotted, using Hofstede's scores for each set of value dimensions for each country. Luxembourg was not in Hofstede's original work but was scored by the authors and has been added to achieve the purpose of this paper.

The striking feature of Exhibit 3 is that the 12 EU countries fall in three different quadrants. Four countries are located in the village market quadrant—Denmark, Ireland, the United Kingdom, and the Netherlands. In this organizational model there is an absence of a decisive hierarchy and problems are solved through negotiation. The village market model leads to accounting systems which are flexible and ad hoc.

Germany, Luxembourg, and Italy (just barely) lie in the well-oiled machine quadrant where financial reporting requires adherence to an established system of rules and a need for predictability. Financial statement preparation is a matter of complying with the established rules.

The remaining five EU countries, Spain, France, Belgium, Portugal, and Greece are in the pyramid of people quadrant. The pyramid is representative of a hierarchical structure held together by rules. Accounting systems are detailed and used to justify the decisions of those in authority and command. Once again, financial statement preparation is a process of complying with the law.

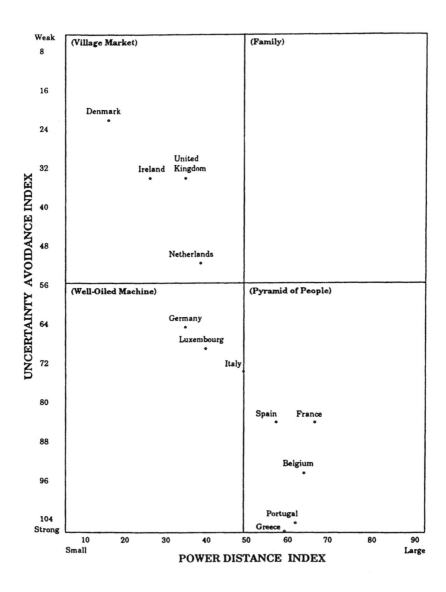

Source: Adapted from Hofstede (1980).

Exhibit 3.   Power Distance by Uncertainty
Avoidance Plot for EU Countries

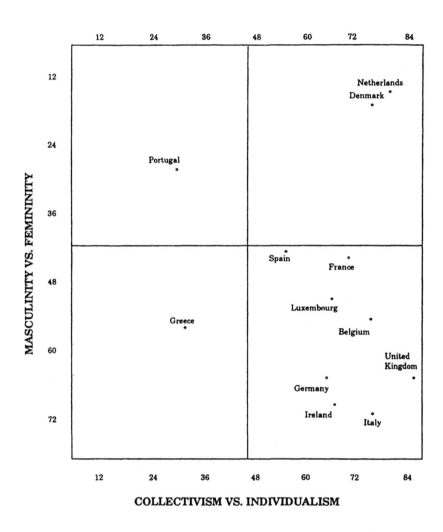

Source: Adapted from Hofstede (1980).

Exhibit 4. Masculinity by Individualism
Plot for EU Countries

Overall, from Exhibit 3 one can see a stark portrayal of fundamental national differences in the cultural values of Power Distance and Uncertainty Avoidance across the 12 EU countries. These differences greatly influence the accounting systems in each of these countries. In part due to these cultural value differences, the EU Directives had to take a mutual recognition approach to developing financial reporting practices, an approach which resulted in bounded diversity, not comparability of these practices.

Exhibit 4 provides a plot of the cultural dimensions of Masculinity and Individualism, attitudes that help describe how individuals behave within an organization. The majority of EU countries scored high on Individualism, indicating that the degree of interdependence among individuals in these countries is relatively low as compared with Spain, Portugal, and Greece. However, the countries varied greatly on the Masculinity value dimension.

It would be difficult to prove that there is a direct correlation between cultural attitudes and accounting policy formation and practice. Hofstede's work did not attempt to be so specific. He provided a set of values for various cultures and laid the foundation for linking these values with financial reporting practices of individual companies. This linkage was further explored by Gray (1988).

Accounting Values

Gray (1988) extended Hofstede's work to link societal values with accounting values. He argued that accountants' value systems do not develop independently but are derived from and relate to underlying societal values, especially the work-related values studied by Hofstede. Thus, Gray hypothesized that cultural values will lead to accounting values at the subcultural level and will influence accounting systems. By reviewing accounting literature and practice, he derived the following accounting value dimensions: Professionalism versus Statutory Control; Uniformity versus Flexibility; Conservatism versus Optimism; and Secrecy versus Transparency.[4]

Gray then offered a series of hypotheses that relate accounting values to Hofstede's cultural values. The hypothesized relationships are as follows:

High individualism
Weak uncertainty avoidance   lead to Professionalism
Small power distance

Strong uncertainty avoidance
Large power distance   lead to Uniformity
Low individualism

Strong uncertainty avoidance
Low individualism   lead to Conservatism
Low masculinity

Strong uncertainty avoidance
Large power distance   lead to Secrecy
Low individualism
Low masculinity

Professionalism is operationalized by exhibiting a preference for and trust in independent professional judgment. A culture that values Individualism generally prefers and allows substance over form and true and fair view to be characteristics that make information useful. This is consistent with weak Uncertainty Avoidance where practice has priority over rigid rules.

Uniformity of accounting rules and financial reporting practices is valued by cultures that are uncomfortable with uncertainty and ambiguity. These cultural environments have little tolerance for flexibility, professional judgment (Individualism), or reporting a true and fair view. They are comfortable with complying with the law.

The generally accepted accounting principle of conservatism results in choosing the accounting method that has the least favorable impact on net income, adopting the most cautious approach of coping with the uncertainty of the future (e.g., strong Uncertainty Avoidance). It has been shown (Gray 1980) that those countries whose financial reporting practices are most influenced by tax laws, lack of strong capital market development, and relative nonexistence of external users report the most conservative net incomes (e.g., Germany and France).

Secrecy can be linked with an unwillingness to disclose information. Or, it may be that it is unnecessary to disclose the information due to the identified user (e.g., banks or governments). External communication of information is less needed and less desirable as a culture moves toward the value of Collectivism and away from Individualism. Cultures that value Collectivism see themselves as belonging to a group or a firm rather than as individuals. In order to preserve the security of the firm (strong Uncertainty Avoidance), it becomes necessary to restrict the amount of information that is available to those who are unrelated and, perhaps, dangerous. Large Power Distance is related to this restriction of information as Secrecy is necessary to preserve the power inequities. Societies that rank lower in Masculinity care more about people and the environment and find it easier to disclose socially responsible information leading to Transparency, more disclosure.

Professionalism and Uniformity seem to be most relevant to the professional or statutory authority for accounting systems and their enforcement, the regulatory environment. Conservatism and Secrecy are relevant to measurement and disclosure practices.

A plot of each of these two relationships for the 12 EU countries is found in Exhibit 5. These plots are replications of Gray's work where he combines appropriate values, provides a classification of culture areas hypothesized on a judgmental basis, and plots the relationship between the two cultural values. While Gray's plots are not numerically based, his judgments rely on the correlations between value dimensions and country clusters identified by Hofstede who used numerical scores for the work-related values.

This classification of EU countries is quite revealing. It shows that Portugal, Greece, Belgium, France, Italy, Spain, and Luxembourg have systems that favor the accounting value of Uniformity. Germany also falls closer to Uniformity than Flexibility. This placement for Germany is consistent with higher Hofstede scores in Uncertainty Avoidance and the influence of tax law on financial reporting in these countries. In Denmark, the Netherlands, the United Kingdom, and Ireland the accounting systems allow for more Flexibility.

The EU countries do not exhibit such a wide range along the Professionalism versus Statutory Control accounting value. Portugal and Greece scored low on Hofstede's cultural value dimension of Individualism from which it follows that they exhibit more Statutory Control in their regulatory environments. The remaining EU countries' accounting systems are influenced to some degree by Professionalism. The influence of Professionalism that has resulted from the adoption of the EU Directives is a change for many countries and is the result of their struggle to incorporate the concept of true and fair view into their financial reporting systems. Later in the paper the authors refer to this influence as Europeanism.

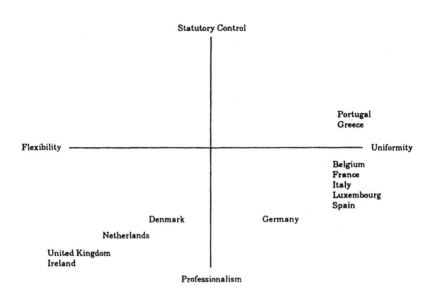

MEASUREMENT AND DISCLOSURE PLOT FOR EU COUNTRIES

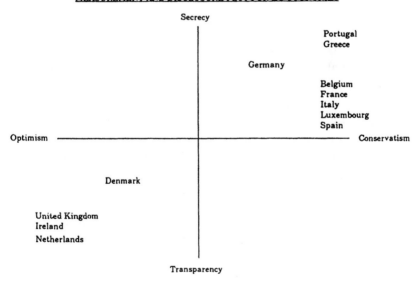

Source: Adapted from Gray (1980).

Exhibit 5. Accounting Value Plots for EU Countries

The measurement and disclosure plot also shows a contrast between those EU countries that embrace the accounting values of Secrecy and Conservatism and those that embrace Transparency and Optimism. This extreme and serious contrast has major implications for achieving comparability of measurement and disclosure reporting practices among EU countries. Conservatism has a direct influence on the valuation of assets and the measurement of net income. Portugal, Greece, Belgium, France, Italy, Luxembourg, Spain, and to a lesser extent Germany, all have financial reporting practices that are influenced by and closely linked to their tax laws. In these countries, tax laws contain a variety of provisions which result in smaller taxable income, including the immediate expensing of asset purchases.

The 12 EU countries are quite divided on disclosure. Eight exhibit more Secrecy, the traditional reluctance to disclose information. Four exhibit Transparency, the traditional willingness to disclose information. The authors anticipate that the 4th Directive will result in less contrast on this Secrecy dimension over time. It is much easier to disclose more information than to change the very measurement practices upon which your system is based.

Modes of Regulation

Puxty et al. (1987) originally analyzed accounting regulation in the United Kingdom, the United States, Germany, and Sweden by observing the influence of market forces (dispersed competition), bureaucratic controls (hierarchical control), and communitarian ideals (spontaneous solidarity) on accounting regulation and accounting practices. They labeled these principles as Market, State, and Community, respectively. Accounting regulation and the resultant required and voluntary financial reporting practices are influenced by a mix of all three of these principles.

Information is provided as the capital market demands it with the result being more disclosure. The state is involved when accounting legislation is passed in the form of Companies Acts, commercial codes, or stock market regulations. The Community provides the pressure for honesty, reliability, and integrity in financial reporting. The Community legitimizes the profession and the product.

Puxty et al. identified four modes of regulation, liberalism, legalism, associationism, and corporatism.[5] This paper adds the

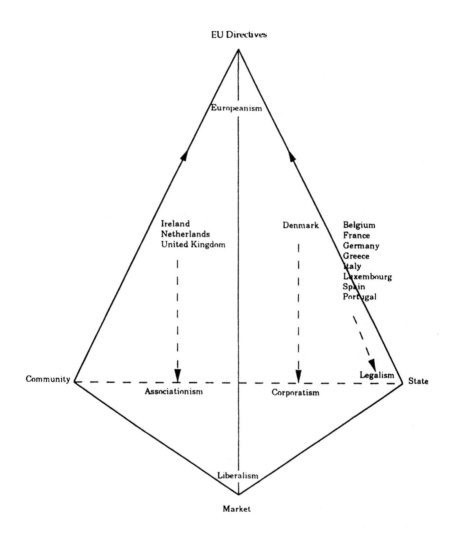

Source: Adapted from Puxty (1987).

Exhibit 6. Degree of Influence of Market, State, Community, and
EU Directives on Accounting Practices of the EU Countries

dimension of the Directives (Europeanism) to the Puxty et al. model, recognizing the influence of the Directives on the financial reporting practices of EU countries. In addition, after reviewing prior research and existing descriptions of the regulatory environments in the 12 EU countries, the authors have placed the countries within Puxty's modes of regulation triangle (anchored by Community, Market, and State) and redrawn the triangle to capture the dimension of Europeanism. Exhibit 6 provides a picture of this adaptation. Europeanism represents the extent to which a country agrees to compromise a part of its accounting culture, accounting values, and its accounting regulatory environment in order to implement the Directives. This revised model allows us to include the "true and fair view" and "standardized formats of reporting" required by the Directives.

The Directives can be considered as both an external (the 12 countries have different points of view but they all agreed to adopt the Directives) and an internal influence (each country must make changes to move toward comparability). So, a country like Germany moves a little away from Legalism under the influence of Europeanism and the Directives. The United Kingdom moves a little away from Associationism while each of the 12 countries moves a little more toward each other.

The models we have discussed, Hofstede, Gray, and Puxty et al., show that there are fundamental, inherent differences in cultural values, in accounting values, and in modes of accounting regulation across the 12 European Union countries that are striving to achieve comparability of financial reporting practices. It is important to recognize these fundamental differences and it is important to understand the influence of Europeanism on valuation, measurement, and disclosure practices. Although there are inherently sturdy generally accepted (acceptable) accounting principles in each country that are in transition, it is important to remember that this transition will take time and that comparability will not be achieved quickly, if at all.

Given that there are fundamental differences in culture and accounting values, as well as in regulatory environments, we can hypothesize that differences in valuation, measurement, and disclosure practices across the 12 EU countries will persist. And, given that the EU Directives allow for choice (diversity) in valuation principles and reporting practices, the previous hypothesis is even more likely to be proven true.

VALUATION, MEASUREMENT AND DISCLOSURE PRACTICES

The differences in measurement and disclosure values among the EU countries can be assessed by examining the plots based on Gray's (1988) accounting values. The plot of measurement and disclosure values in Exhibit 5 provides evidence of a revealing, but not surprising, split among the EU countries. Most of the Continental European countries score high in conservatism which implies a strongly conservative approach to valuation and measurement. This contrasts with the less conservative approach of Denmark, the United Kingdom, Ireland, and the Netherlands. Gray (1980) has shown that differences in accounting practices affect reported corporate profits in France, Germany, and the United Kingdom. Gray's results suggest that the measurement of profits is correlated with national characteristics. Companies in France and Germany exhibit measurement behavior that leads to the reporting of conservative profit numbers. The influence of tax law and government, combined with less development in capital markets, accounts for much of the conservative behavior of the Continental European countries.

This persistent contrast between optimism and conservatism continues to render the reported financial results of companies noncomparable across countries. While the Directives strived to achieve more comparability, they were more concerned with broad issues rather than detailed rules or the elimination of choices. This approach allowed countries to accept, adopt, and implement the Directives, but it left many of the important accounting details unresolved. It allowed for choice and flexibility, which as the history of the International Accounting Standards Committee has shown, lead to a lack of expected comparability (Wallace 1990; Purvis, Gernon, and Diamond 1991).

Increased disclosure can help to solve the problems encountered when dealing with diverse sets of generally accepted accounting principles. Historically, the extensiveness of disclosure, or lack thereof, has been decided by the needs of the users of the information. More disclosure has been the norm in the United Kingdom, Ireland, and the Netherlands. This phenomenon seems to be related to the development of the equity market and the providers of finance being investors. The majority of EU countries have relied on bank and government financing which has led to a more conservative, and less transparent mode of accounting. This mode supplied the users' needs. As users' needs change additional information will be demanded and provided by additional disclosures. This will be an evolving process throughout Continental Europe and one that will occur at various time intervals.

SUMMARY, CONCLUSIONS, AND FUTURE RESEARCH

Across Europe, users of financial accounting information are changing and their needs are changing. Raising capital in equity markets has more general appeal now and the disclosure policies of companies are in transition due to their attempt to satisfy users' needs. Increased comparability of financial reporting could facilitate the formation and movement of capital within the European Union, but this can only be determined with further research.

One of the objectives of the Directives of the European Commission was the reduction in financial reporting differences across the 12 EU countries. Given the diversity of cultural and accounting values as well as modes of regulation across the EU countries, the Directives could not achieve the state of comparability in financial accounting practices. However, the Directives were able to set minimum standards which reduced the number of allowable options thus resulting in bounded diversity. Additional disclosure, completeness, and comparability will be forthcoming as they are demanded by the various users of the information. Comparability remains an elusive goal for very supportable reasons and would only be achieved if choice and flexibility could be removed from the Directives. However, removing choice and flexibility could result in making the reported financial information less useful to the domestic user. Over time the needs of the domestic user may evolve to a point where they are similar and compatible with the needs of the international user, but this remains to be seen. These issues represent ideas for further research.

Prior studies have shown that there are cultural differences throughout the EU countries that lead to different modes of regulation, different financial reporting practices, and different users of accounting information. Differences in financial reporting practices persist after the Directives because the politically palatable mutual recognition approach required flexibility and acceptance of a country's domestic peculiarities. The Directives moved the EU countries toward a common financial reporting goal. Europeanism has made financial reporting more consistent and complete, but movement into the state of comparability remains elusive and, perhaps, ill-advised. Change takes time, and with time, the choices allowed by the Directives can be lessened if this appears to be a worthy goal. This would result in a reduction of accounting differences and a move toward the European Union goal of comparability.

ACKNOWLEDGMENT

We appreciate helpful comments and suggestions from workshop participants at the University of Glasgow and the University of Wales College of Cardiff. In addition, we would like to thank two anonymous reviewers and Gary Previts, Editor, for their helpful comments. We are also grateful for financial support provided by the University of Oregon and the University of Alabama.

NOTES

1. At the time of this analysis there were only 12 member countries by of the EU.
2. At a later date, supplementary data for another 10 countries became available (Hofstede 1983b, 80).
3. The meaning of each dimension is well described by Hofstede (1984, 83-4):

Individualism versus Collectivism
Individualism stands for a preference for a loosely knit social framework in society wherein individuals are supposed to take care of themselves and their immediate families only. Its opposite, Collectivism, stands for a preference for a tightly knit social framework in which individuals can expect their relatives, clan, or other in-group to look after them in exchange for unquestioning loyalty (it will be clear that the word "collectivism" is not used here to describe any particular political system). The fundamental issue addressed by this dimension is the degree of interdependence a society maintains among individuals. It relates to people's self-concept: "I" or "we."

Large versus Small Power Distance
Power Distance is the extent to which the members of a society accept that power in institutions and organizations is distributed unequally. This affects the behavior of the less powerful as well as the more powerful members of society. People in Large Power Distance societies accept a hierarchical order in which everybody has a place that needs no further justification. People in Small Power Distance societies strive for power equalization and demand justification for power inequalities. The

fundamental issue addressed by this dimension is how a society handles inequalities among people when they occur. This has obvious consequences for the way people build their institutions and organizations.

Strong versus Weak Uncertainty Avoidance

Uncertainty Avoidance is the degree to which the members of a society feel uncomfortable with uncertainty and ambiguity. This feeling leads them to beliefs promising certainty and to maintaining institutions protecting conformity. Strong Uncertainty Avoidance societies maintain rigid codes of belief and behavior and are intolerant towards deviant persons and ideas. Weak Uncertainty Avoidance societies maintain a more relaxed atmosphere in which practice counts more than principles and deviation is more easily tolerated. The fundamental issue addressed by this dimension is how a society reacts on the fact that time only runs one way and that the future is unknown: whether it tries to control the future or to let it happen. Like Power Distance, Uncertainty Avoidance has consequences for the way people build their institutions and organizations.

Masculinity versus Femininity

Masculinity stands for a preference in society for achievement, heroism, assertiveness, and material success. Its opposite, Femininity, stands for a preference for relationships, modesty, caring for the weak, and the quality of life. The fundamental issue addressed by this dimension is the way in which a society allocates social (as opposed to biological) roles to the sexes.

4. *Professionalism versus Statutory Control*—a preference for the exercise of individual professional judgment and the maintenance of professional self-regulation as opposed to compliance with prescriptive legal requirements and statutory control.

Uniformity versus Flexibility—a preference for the enforcement of uniform accounting practices between companies and for the consistent use of such practices over time as opposed to flexibility in accordance with the perceived circumstances of individual companies.

Conservatism versus Optimism—a preference for a cautious approach to measurement so as to cope with the uncertainty of future events as opposed to a more optimistic, laissez-faire, risk-taking approach.

Secrecy versus Transparency—a preference for confidentiality and the restriction of disclosure of information about the business only to those who are closely involved with its management and financing as opposed to a more transparent, open and publicly accountable approach (Gray 1988, 8).

5. *Liberalism*—regulation is provided exclusively by the discipline of Market principles. Information is provided if it is found to be commercially demanded.

Legalism—relies upon the unreserved application of State principles. Behavior is sanctioned if it follows the letter of the law. Germany is an example.

Associationism—a mixed model where regulation is accomplished through the development of organizations that are formed to represent and advance the interests of their members (i.e., professional accounting bodies). The United Kingdom exemplifies associationism.

Corporatism—a mixed mode where the State incorporates organized interest groups into its own centralized, hierarchical system of regulation. The State simultaneously recognizes its dependence these associations and seeks to use them as an instrument in the pursuit and legitimation of its policies. Sweden provides an example of Corporatism (Puxty et al. 1987, 282).

REFERENCES

Biddle, G.C., and S.M. Saudagaran. 1989. The effects of financial disclosure levels on firms' choices among alternative foreign stock exchange listings. *Journal of International Financial Management and Accounting* 1(Spring): 55-87.

Choi, F.D.S., and R.M. Levich. 1990. *The Capital Market Effects of International Accounting Diversity. Homewood, IL: Dow Jones-Irwin.*

Cooke, T.E. 1991. *The evolution of financial reporting in Japan: A shame culture perspective. Accounting, Business and Financial History* 1(3): 251-277.

Farmer, R., and B. Richman. 1966. *International Business: An Operational Theory.* Homewood, IL: Irwin.

Gray, S.J. 1988. Towards a theory of cultural influence on the development of accounting systems internationally. *ABACUS* 24(1): 1-15.

_____. 1980. The impact of international accounting differences from a security-analysis perspective; some European evidence. *Journal of Accounting Research* (Spring): 64-76.

Harrison, G.L., and J.L. McKinnon. 1986. Culture and accounting change: A new perspective on corporate reporting regulation and accounting policy formulation. *Accounting, Organization and Society* 11(3): 233-252.

Haskins, M. 1991. International Financial Reporting. Unpublished working paper. Charlottsville, VA: Darden Graduate Business School.

Hofstede, G. 1980. *Culture's Consequences.* Beverly Hills, CA: Sage.

_____. 1983a. Dimensions of national cultures in fifty countries and three regions. In *Explications in Cross-Cultural Psychology*, eds. J.B. Deregowski, S. Dziurawiec and R. Annis, 335-355. Lisse: Swets and Zeitlinger.

_____. 1983b. The cultural relativity of organizational practices and theories. *Journal of International Business Studies* (Fall): 75-89.

_____. 1984. Cultural dimensions in management and planning. *Asia Pacific Journal of Management* (January): 81-99.

_____. 1987. The cultural context of accounting. In *Accounting and Culture*, ed. B.E. Cushing, 1-12. Sarasota, FL: American Accounting Association.

Kirsch, R.J. 1994. Toward a global reporting model: Culture and disclosure in selected capital markets. In *Research in Accounting Regulation*, Vol. 8, ed. G.J. Previts, 71-109. Greenwich, CT: JAI Press.

McKinnon, J.L., and G.L. Harrison. 1985. Cultural influence on corporate and governmental involvement in accounting policy determination in Japan. *Journal of Accounting and Public Policy* (Fall): 201-223.

Mueller, G.G. 1967. *International Accounting.* New York: Macmillan.

_____. 1991. 1992 and harmonization efforts in the E.C. In *Handbook of International Accounting*, ed. F.D.S. Choi, 1-30. New York: Wiley.

Nair, R.D., and W.G. Frank. 1980. The impact of disclosure and measurement
    practices on international accounting classifications. *The Accounting Review*
    (July): 426-450.
Nobes, C.W. 1983. A judgmental international classification of financial reporting
    practices. *Journal of Business Finance and Accounting* (Spring): 1-19.
Purvis, S.E.C., H. Gernon, and M. Diamond. 1991. The IASC and its comparability
    project. *Accounting Horizons* 5(2): 25-44.
Puxty, A.G., H.C. Willmott, D.J. Cooper, and T. Lowe. 1987. Modes of regulation
    in advanced capitalism: locating accountancy in four countries. *Accounting,
    Organizations and Society* 12(3): 273-291.
Radebaugh, L.H., and S.J. Gray. 1993. *International Accounting and Multinational
    Enterprises.* New York: Wiley.
Tay, J.S.W., and R.H. Parker. 1990. Measuring international harmonization and
    standardization. *ABACUS* 26(1): 71-88.
Van Hulle, K. 1989a. The EC experience of harmonization—Part I. *Accountancy*
    (September): 86-91.
_____. 1989b. The EC experience and harmonization—Part II. *Accountancy*
    (October): 96-99.
Wallace, R.S.O. 1990. Survival strategies of a global organization. The case of the
    IASC. *Accounting Horizons* 4(2): 1-22.

SECURITY PRICE RESPONSE ASSOCIATED WITH THE ACCOUNTING REGULATION OF PURCHASE COMBINATIONS INCREASING GOODWILL

Kathleen Blackburn Hethcox

ABSTRACT

This study examines whether the popular belief that the large decrease in reported earnings from the required amortization of purchased goodwill is supported empirically. Specifically, is the purchase transaction that increases goodwill associated with a decrease in the security prices of acquiring companies? Additionally, this study investigates whether the magnitude of the goodwill change is negatively related to the magnitude of the stock price reaction. Employing a sample of 88 firms reporting increases in goodwill

Research in Accounting Regulation, Volume 9, pages 49-62.
Copyright © 1995 by JAI Press Inc.
All rights of reproduction in any form reserved.
ISBN: 1-55938-883-8

from 1988 to 1989 and using CRSP and Compustat data, the cumulative abnormal returns (CARs) of the firms over a 24-month test period were analyzed. The results indicate a significant decrease in stock price associated with mergers that were accounted for by the purchase method and had goodwill increases. Further, the greater the increase in goodwill, the greater the stock price decline.

With the new wave of mergers and acquisitions occurring in the 1980s and 1990s, purchased goodwill has received much negative attention in the financial press. Stakeholders purport that the large decrease in reported earnings from GAAP's required goodwill amortization lowers the stock market's valuation of the company (e.g., see Linden 1990). Accountants have appealed to the FASB to review business combination accounting, asserting that it is affecting the international competitiveness of U.S. companies that bid against foreign firms (Dieter 1989). During the time these complaints have been made, there have been no direct cash flow effects from goodwill amortization, because the IRS has not allowed the amortization for tax purposes.[1] It is not immediately obvious that the opinions expressed in the financial press are correct because accounting researchers have investigated the information content of accounting methods, with most finding that accounting manipulations that are not accompanied by real economic impacts have no statistically significant effect on security prices (Lev and Ohlson 1982).

This study examines whether the popular beliefs are supported empirically; that is, is the purchase transaction that increases goodwill associated with a decrease in the security prices of acquiring companies? Further, the study investigates whether the magnitude of the goodwill change is related to the magnitude of the stock price reaction.

Over the years, accounting for business combinations has been controversial. *APB Opinion Nos. 16* and *17* were adopted in 1970 as a compromise. The rules for pooling and purchase are outlined in *APB Opinion No. 16.* In the pooling-of-interests method, the book value assets of the two merging companies are combined for the resulting entity. For a purchase transaction, in addition to the recording of goodwill, assets of the acquired company are valued at market, thus frequently increasing the depreciation base for fixed assets. The resulting increased depreciation charges can decrease reported earnings. In addition, *APB Opinion No. 17* requires the amortization of goodwill, which further decreases reported earnings. Consequently, since 1970, pooling accounting has been the method that should allow greater reported earnings.

Because the Internal Revenue Code has not allowed for the amortization of goodwill for tax purposes in years past, some might argue that managers have not been that concerned about goodwill amortization because they know it has no actual cash flow effect. However, results from the pooling versus purchase method literature have indicated otherwise. Assuming that managers seek to maximize reported income, Gagnon (1967) demonstrated managers choose pooling over purchase if the difference between the purchase price and the asset values is substantial.

Robinson and Shane (1990) posited that the average benefit derived from the accounting method is greater for pooling than for purchases. They argued that because it is relatively easy to structure a purchase but more difficult to structure a pooling, economic benefit must be derived from the accounting effects of pooling. Their results indicate that managers are willing to pay more in order to structure a merger as a pooling, presumably because pooling usually results in higher earnings (Robinson and Shane 1990).

Investigating the association between purchase method accounting and the security price response provides the accounting community more information with which to evaluate the efficacy of current GAAP for business combinations. If empirical results support the popular beliefs, implications are that the security market can react to acounting transactions with no direct cash flow effects and that perhaps market valuation is adversely affected by financial reporting regulations.

In the remainder of this paper, the research hypotheses and design will be described. Then, the results will be presented, followed by some summary remarks.

RESEARCH HYPOTHESES AND DESIGN

Hypotheses

When an acquisition is accounted for as a purchase, investors must value the combined entity. Details of the purchase price and the goodwill amount with its amortization period are publicly available in acquisition agreements, proxy statements, and annual reports. The first research hypothesis tests the popular belief that goodwill is associated with a decrease in firm valuation.

> **Hypothesis 1**:   Firms involved in mergers that use the purchase method of accounting, which results in a material increase in goodwill, experience a decrease in stock price.

Firms that use a purchase combination report differing amounts of goodwill depending on the amount of excess of the purchase price over the net asset values of the target firm. If abnormal returns are associated with the amount of goodwill recorded, then varying impacts might be observed among firms depending on the amounts of recorded goodwill. The second hypothesis tested in this study is as follows.

> **Hypothesis 2**:   For firms involved in a merger accounted for by the purchase method, those with a larger percentage increase in goodwill experience a greater decrease in stock price.

Rejection of the null hypothesis would indicate a negative association between the goodwill change and the market valuation of the firm. This finding would further support the popular assertions about the negative impacts of goodwill amortization.

Data and Sample Selection

The study is structured as an association study to allow the testing of stock price reaction without identifying the exact merger date. Mergers with goodwill increases are identified from Compustat which began reporting goodwill as a separate item in 1988. Consequently, changes in goodwill cannot be determined from the Compustat file until 1989. The test period of 1989-1990 was used, as it provided 12 months of returns before and after the date at which the goodwill increase from a merger first appeared in the annual financial statements. The sample firms were selected from the 1990 Compustat Annual Industrial File and from the 1990 CRSP Monthly Returns File according to the following criteria:

1. Goodwill increased from 1988 to 1989.
2. The firm had a December fiscal year-end.
3. The firm did not record a merger by pooling in the same year.
4. The goodwill was significantly large relative to the acquiring firm.
5. Monthly return data were available for the firm for 1984-1990.

Criterion 4 eliminated small and immaterial acquisitions. To implement this test, a goodwill change (GWC) was computed for each firm. The formula used for computing the change in goodwill is as follows:

$$GWC = \frac{Goodwill_{89} - Goodwill_{88} + Amortization\ of\ Goodwill_{89}}{Market\ Value\ of\ Equity_{88}}$$

Firms with a GWC less than .01 were eliminated from the sample.

Previous research has documented differences in price reactions to mergers depending on the method of payment for the merger, cash or stock (Travlos 1987), and tax status of the merger, taxable or nontaxable (Brown and Ryngaert 1991). These closely related variables were controlled by the sample selection. *APB Opinion No. 16* requires that any acquisition involving cash be accounted for as a purchase combination; thus, the sample contained only acquisitions obtained partially or totally with cash. A review of events for each sample firm in the *Wall Street Journal Index* for 1989 revealed all sample acquisitions were purchases made with a majority of cash, except one.[2]

If greater than 50% of the purchase price is paid with cash, a merger is classified as taxable. IRS approval is required to classify a merger as nontaxable in this case. Because all but one merger in the sample was for a majority of cash, it was assumed these were taxable mergers.[3]

The initial sample consisted of 317 firms that had increases in goodwill in 1989. Of these, 116 firms did not have a December fiscal year-end, five firms had concurrent poolings, 13 firms were missing share or price data, and 63 firms had incomplete return data. Thirty-two firms with GWC less than .01 were eliminated from the sample. The final test sample consisted of 88 firms. Table 1 illustrates the sample selection process.

Table 1. Test Sample After Applying
Selection Criteria

1.	Goodwill in 1989 > Goodwill in 1988	317
2.	Non-December fiscal year-end firms	(116)
3.	Firms with pooling during 1989	(5)
4.	Firms missing share or price data	(13)
5.	Complete return data for 1984-1990 unavailable	(63)
6.	Firms with goodwill change < 1%	(32)
	Total sample firms	*88*

Expectations Model and Abnormal Returns

The test period was defined as January 1989 through December 1990. The market model developed by Sharpe and others was employed to compute estimated returns from 1984-1988—the 60-month period prior to the test period. Abnormal returns for each firm for each of the 24 test period months were estimated by subtracting the estimated return from the actual return for the firm in the specific month. In addition, cumulative abnormal returns (CARs) were calculated for each individual firm, and average CARs were calculated for each sample tested.

In association studies, abnormal return behavior in months surrounding a critical event is examined. In this study, earnings announcements were used as a surrogate for the publicly available information about the merger agreement and the recorded goodwill.[4] Because preliminary earnings information usually is published along with other performance information prior to the end of the fiscal year, it is reasonable to expect that much of the price reaction to earnings will be incorporated by year-end. Consequently, following the approach of Beaver, Lambert, and Morse (1980) and Rayburn (1986), December 31, 1989, was identified as the critical date—the first annual earnings period ending after a merger that results in an increase of recorded goodwill. Because all firms have a fiscal year-end of December 31, earnings announcements should be closely aligned even if actual announcement dates differ. If no earnings information was announced until after the year-end date, the regression coefficients were biased toward zero.

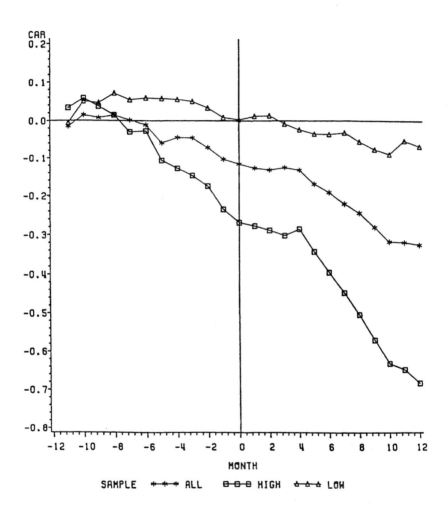

Note: 88 total sample firms, 18 each in low and high quintiles from firms ranked by percentage change in goodwill.

Figure 1. Average Cars for Three Firm Groups
Month 0 = December 1989

Table 2. Average Cumulative Abnormal Returns
(CARs) by Sample Groups
Month 0 = December, 1989

	Sample Groups		
Month	ALL $n = 88$	HIGH $n = 18$	LOW $n = 18$
− 11	− .0146	.0330	− .0055
− 10	.0145	.0576	.0499
− 9	.0076	.0368	.0459
− 8	.0147	.0146	.0708
− 7	.0012	− .0298	.0539
− 6	− .0110	− .0272	.0588
− 5	− .0591*	− .1065	.0577*
− 4	− .0445*	− .1275	.0551*
− 3	− .0453*	− .1469	.0501*
− 2	− .0717**	− .1750	.0332**
− 1	− .1031**	− .2347	.0085**
0	− .1156**	− .2684	.0029**
+ 1	− .1265**	− .2771	.0121**
+ 2	− .1314**	− .2880	.0138**
+ 3	− .1251**	− .3011	− .0082*
+ 4	− .1319**	− .2845	− .0236*
+ 5	− .1692**	− .3421	− .0350*
+ 6	− .1909**	− .3948	− .0362*
+ 7	− .2205**	− .4482	− .0320*
+ 8	− .2440**	− .5048	− .0562*
+ 9	− .2806**	− .5709	− .0770*
+ 10	− .3171**	− .6326	− .0895**
+ 11	− .3187**	− .6485	− .0533**
+ 12	− .3249**	− .6838	− .0679*

Notes: HIGH and LOW groups formed from upper and lower quintiles of firms ranked
on percentage change in goodwill. (Percentage goodwill change = change in
goodwill/market value of equity.)

One-tailed t-tests for ALL sample:
** significantly less than zero at < .01 level
* significantly less than zero at < .05 level

One-tailed t-tests for HIGH vs. LOW samples:
** significantly lower means for HIGH sample at < .01 level
* significantly lower means for HIGH sample at < .05 level

RESULTS

Hypothesis 1

In Figure 1, the ALL line illustrates the average $CARs$ for the total
sample of 88 firms. Table 2 reports the 24-month test period average

*CAR*s for the total sample in the second column. A decrease in abnormal returns associated with merger can be observed.

One-tailed *t*-tests were performed to examine whether these *CAR*s were less than zero, and from month –5 through month +12, the average *CAR*s were significantly less than zero. From month –2 through month +12, these *CAR*s were significant at less than the .01 level. Thus, these results support the first hypothesis indicating that a significant decrease in stock price is associated with mergers accounted for by the purchase method with goodwill increases.

Hypothesis 2

To test for differences between groups with a relatively large percentage of goodwill and a relatively small percentage, the 88 firms were ranked according to *GWC*. The upper quintile of 18 firms (HIGH) consists of firms with relatively large goodwill changes ranging from 0.158 to 1.522. The lower quintile of 18 firms (LOW) has relatively small goodwill changes ranging from 0.010 to 0.022. The *CAR*s for these groups are illustrated in Figure 1 and listed in Table 2.

One-tailed *t*-tests for the difference in means between the HIGH and LOW groups were performed. From month –5 through month +12, the HIGH group *CAR*s were significantly lower than the LOW group *CAR*s at less than the .05 level. At seven months the HIGH CARs were significantly lower at the .01 level. These results support the second hypothesis. Consequently, there is evidence that a negative association exists between the percentage change in firms' goodwill and their stock prices, and the greater the increase in goodwill, the greater the stock price decline.[5]

Cross-Sectional Regressions

Cross-sectional regression was employed to test the overall significance of the goodwill change on the cumulative abnormal returns. Independent variables included in the model were the change in goodwill and the change in earnings per share. Previous studies have shown that changes in earnings are associated with abnormal returns (Ball and Brown 1968; Beaver 1968; Beaver, Clarke, and Wright 1979). To make the data comparable across firms and account for size differences, each variable was deflated by the market value of equity. The variables were calculated from Compustat data as follows:

$$\Delta GW = (GW89 - GW88 + AMORT89) \,/\, MKTVAL88$$

where: $GW88$ and $GW89$ = balance sheet goodwill in 88 and 89, respectively

$AMORT89$ = amortization of intangibles in 89, times the ratio of goodwill in 89 to intangibles in 89

$MKTVAL88$ = common shares outstanding in 88, times price at close of fiscal year 88

$$\Delta EPS = (EPS89 - EPS88) \,/\, MKTVAL88$$

where: $EPS88$ and $EPS89$ = fully diluted earnings per share in 88 and 89

Spearman Correlations were computed to test the fit of the proposed variables. ΔGW and ΔEPS both significantly correlate with CAR at less than the 0.01 level (correlation coefficients = $^-0.309$ and 0.401, respectively) and in the hypothesized direction. However, the independent variables, ΔGW and ΔEPS do not significantly correlate with each other, thus avoiding problems of multicollinearity in the model. The resulting regression equation is:

$$CAR_t = b_0 + b_1 \frac{\Delta GW_t}{MKTVAL_{t-1}} + b_2 \frac{\Delta EPS_t}{MKTVAL_{t-1}} + e_t$$

The subscript i is omitted for each individual firm observation; b_1 is predicted to be negative because it is hypothesized that as goodwill increases, stock prices decrease; and b_2 is predicted to be positive because increases in earnings are viewed positively by the market.

Table 3 presents the regression results for the total sample of 88 individual firm observations. The overall model is significant at less than the 0.0009 level. The coefficient of ΔGW is significant and negative at less than the 0.025 level (one-tailed t-test). The coefficient of ΔEPS is significant and positive at less than the 0.0005 level (one-tailed t-test). This result indicates that the abnormal returns are negatively related to the goodwill change, thus lending additional support to the first and second hypotheses.

Table 3.   Cross-Sectional Regression of
Individual CARs for the Total Sample
$n = 88$
Model: $CAR_{it} = b_0 + b_1 \Delta GW_{it} + b_2 \Delta EPS_{it} + e_{it}$

	Intercept	ΔGW	ΔEPS
Expected sign		–	+
Coefficient	– 0.250	– 0.652	9.685
t-value	– 3.289	– 1.996	3.375
Significance (one-tailed)	(.0007)	(.0245)	(.0005)

Notes: F-value for overall model = 7.668 (significant at < .0009)
   $R^2 = .1528$
   Adjusted $R^2 = .1329$

SUMMARY AND IMPLICATIONS

This study has tested the relation between stock prices and increases in goodwill that resulted from purchase transactions. The results indicate that stock prices tend to decrease around merger for firms involved in purchases that increase goodwill. Additionally, a differential reaction exists depending on the amount of the goodwill change for the firms; that is, the greater the goodwill change, the more negative the security price response. This finding documents an instance where the security market reacts to an accounting regulation that involves no direct cash flow effect to the firm.

The more negative price reaction to purchases with greater increases in goodwill should be of interest to the professional accounting community as it asks the FASB to review the 20-year-old rules for accounting for business combinations. These rules were adopted before the takeover wave of the 1980s and the increased global competition among firms in buying other companies. Because goodwill is accounted for differently in other countries (and frequently does not decrease aftertax earnings), there are complaints that the United States is on an unlevel playing field when bidding against foreign firms.

In Britain, the preferred treatment of goodwill is to write it off immediately against reserves. This allows British companies to report higher earnings than U.S. companies in the same situation. As a result, the British company can offer a higher purchase price for a common acquisition knowing that its future reported earnings will not be burdened by the higher prices paid. Choi and Lee (1991) found that merger premia associated with United Kingdom acquisitions were consistently higher than those for their U.S. counterparts. Their evidence implies that the GAAP method of accounting for business combinations is affecting the international competitiveness of U.S. firms.

The security price reaction documented in this study is inconsistent with earlier studies reviewed in Lev and Ohlson (1982), which found that accounting choices having no direct cash flow effects did not significantly affect security prices. However, positive accounting theory conjectures that changes in earnings affect the contracting costs of firms, thus indirectly affecting future cash flows. If investors believe reduced earnings will lead to re-contracting costs, then they will lower their valuation of the firm[6]. The expected reduction in future cash flows from possible re-contracting costs might be the reason for the negative price reaction captured in this sample.

In addition, investors may believe that managers make irrational decisions and over-value targets (see Roll 1986). Consequently, their belief that the bidding firm has paid too much for the target causes them to react negatively to the merger. Finally, investors also may adjust financial statement data depending on their own views as to a reasonable life for the recorded goodwill. If investors view the purchased goodwill as having a much shorter life than the allowed 40-year amortization, such as in some high technology mergers, then they may believe that earnings are systematically overstated and reduce their valuation of the firm.[7] Any of the above explanations for the negative price reaction observed in this sample are plausible. However, it is a question for future research to further identify specific phenomena associated with the negative reaction to mergers accounted for as purchase combinations.

NOTES

1. In August 1993, Congress passed a new tax act (OBRA '93) which now allows for the amortization of goodwill for tax purposes over a period not to exceed 15-years. As a tax deduction, goodwill amortization will now have a direct cash flow effect to the firm. However, the 15-year time period requires the write-off of much larger amounts of goodwill than the 40-year amortization period allowed by GAAP.

2. Travlos (1987) and Brown and Ryngaert (1991) classified their cash samples as those where the acquisition was paid with at least 51% cash.

3. No information about the tax status of the one exception acquisition in the sample was found. If this one firm which differs in method of payment and tax status biased the sample, it would be against rejection of the null. In addition, Hayn (1989) found a significant positive relation between the step-up in assets (a tax advantage in purchase combinations) and cumulative abnormal returns of bidding firms involved in taxable mergers. This condition would also bias this sample against rejection of the null. It is noted, however, that the restructuring of the Tax Code in 1986 reduced the benefits from stepped-up assets, and Hayn's sample was for 1970-1985. The sample in this study is after implementation of the 1986 Tax Code change.

4. Prior studies have documented the information content of earnings announcements (e.g., see Ball and Brown 1968; Beaver 1968).

5. The majority of the HIGH group *GWC*s range from 0.158 to 0.568. Two outlier firms have *GWC*s of 0.817 and 1.522. To test for the possibility that these two outliers are affecting the results, all *t*-tests were rerun, after removing these two firms from the total sample and the HIGH group. All results (not reported here) were similar and within the same significance level.

6. Leftwich (1981) and Lys (1984) provide evidence that decreased earnings result in decreased share values when those reduced earnings make debt covenants more binding.

7. Duvall, Jennings, Robinson, and Thompson (1992) discuss these investor differences in valuation and views of goodwill and cite examples from public information.

REFERENCES

Ball, R., and P. Brown. 1968. An empirical evaluation of accounting income numbers. *Journal of Accounting Research* 6(Autumn): 159-178.

Beaver, W.H. 1968. The information content of annual earnings announcements. *Journal of Accounting Research* 6(Supplement): 67-92.

Beaver, W.H., R. Clarke, and W.F. Wright. 1979. The association between unsystematic security returns and the magnitude of earnings forecast errors. *Journal of Accounting Research* 17(Autumn): 316-340.

Beaver, W.H., R. Lambert, and D. Morse. 1980. The information content of security prices. *Journal of Accounting and Economics* 2(March): 3-28.

Brown, D.T., and M.D. Ryngaert. 1991. The mode of acquisition in takeovers: Taxes and asymmetric information. *The Journal of Finance* 46(June): 653-669.

Choi, F.D.S., and C. Lee. 1991. Merger premia and national differences in accounting for goodwill. *Journal of International Financial Management and Accounting* 3: 219-240.

Dieter, R. 1989. Is now the time to revisit accounting for business combinations? *The CPA Journal* 59(July): 44-48.

Duvall, L., R. Jennings, J. Robinson, and R.B. Thompson II. 1992. Can investors unravel the effects of goodwill accounting? *Accounting Horizons* 6(June): 1-14.

Gagnon, J. 1967. Purchase versus pooling of interests: The search for a predictor. *Journal of Accounting Research* (Supplement): 187-204.

Hayn, C. 1989. Tax attributes as determinants of shareholder gains in corporate acquisitions. *Journal of Financial Economics* 23(June): 121-153.

Leftwich, R. 1981. Evidence of the impact of mandatory changes in accounting principles on corporate loan agreements. *Journal of Accounting and Economics* 3(March): 3-36.

Lev, B., and J.A. Ohlson. 1982. Market-based empirical research in accounting: A review, interpretation, and extension. *Journal of Accounting Research* 20(Supplement): 249-322.

Linden, D.W. 1990. The accountants versus the dealmakers. *Forbes,* August 20, p. 84.

Lys, T. 1984. Mandated accounting changes and debt covenants: The case of oil and gas accounting. *Journal of Accounting and Economics* 6(April): 39-65.

Rayburn, J. 1986. The association of operating cash flow and accruals with security returns. *Journal of Accounting Research* 24(Supplement): 112-133.

Robinson, J.R., and P.B. Shane. 1990. Acquisition accounting method and bid premia for target firms. *The Accounting Review* 65(January): 25-48.

Roll, R. 1986. The Hubris hypothesis of corporate takeovers. *Journal of Business* 59(April): 197-216.

Travlos, N.G. 1987. Corporate takeover bids, methods of payment, and bidding firms' stock returns. *The Journal of Finance* 42(September): 943-963.

THE EFFECT OF PEER REVIEW
ON AUDIT ECONOMIES

Gary Giroux, Donald Deis, and Barry Bryan

ABSTRACT

This paper is an extension of the audit economics literature in the public sector, with primary focus on fees and costs of audits, as well as audit quality. A gap in this literature is the relative lack of information on the effects of peer review, which is of major concern in this project. Texas school district information provided by the Texas Education Agency represents a rich database for extensive analysis. Univariate and multivariate analysis of audit fees indicate significance with peer review, suggesting a price premium. However, the lack of significance with fees per hour suggests a more thorough audit and no price premium. Audit quality is strongly associated with peer review and several other factors. Overall, results indicate that peer reviewed audit firms provide higher quality audits with fee premia related to more extensive audit procedures.

Research in Accounting Regulation, Volume 9, pages 63-82.
Copyright © 1995 by JAI Press Inc.
All rights of reproduction in any form reserved.
ISBN: 1-55938-883-8

63

A considerable amount of research has been conducted on the market for financial audits, with particular interest in audit fees and the quality of the audit. The public sector provides an opportunity for further research because of superior data availability from state and federal agencies and audit review programs required by the Single Audit Act and related regulations (Granof 1992). Texas school district information was used for data analysis in this study.

This paper explores the relationship of peer review to traditional audit economic characteristics. Francis, Andrews, and Simon (1990) asked the question: Are peer reviewed firms perceived as quality differentiated auditors? They hypothesize that if there is perceived quality differentiation among firms, then peer reviewed firms should charge higher audit fees. Their testing, based on a multivariate audit fee model, indicated no significant fee differences for peer reviewed firms. They used membership in the American Institute of Certified Public Accountants (AICPA) Division of CPA firms as a surrogate for peer review. Membership in the Division, as well as participation in the peer review program, was voluntary during the period of their study. A limitation of Francis et al. (1990) is that AICPA Division member firms were not necessarily peer reviewed.

This study extends Francis et al. (1990) to the governmental sector where an independent measure of audit quality exists. Two questions are examined: First, are peer reviewed audit firms quality differentiated? Second, do peer reviewed audit firms command audit fee premia? The hypotheses were analyzed using both univariate and multivariate models. During the period under study, peer review was voluntary, with approximately 21% of the Texas school district auditors being peer reviewed. Audit fees were higher for peer reviewed audit firms, but not significantly higher on a fee per hour basis. Audit quality was significantly higher for peer reviewed firms, suggesting that peer reviewed firms provided higher quality audits, with fee premia associated with greater audit effort.

BACKGROUND ON PEER REVIEW

By definition, peer review is a thorough examination of an auditing firm's quality control system. In practice, peer review is a rigorous evaluation of a firm's accounting and auditing practice, carried out by other CPAs under the auspices of the AICPA. Peer review provides assurance to the public that each firm maintains a quality control system appropriate to its practice and that it complies with that system. Peer reviews are carried out following proven comprehensive standards and guidelines. The peer review program was initiated in 1977 with participation on a voluntary basis. The AICPA did not adopt mandatory peer review at that time. However, most members favored it for firms with SEC clients. Wallace (1989) noted that a voluntary program is subject to criticism due to the self-selection bias that can arise because firms most prone to a need for quality control and oversight opt out of the process.

In 1988, the AICPA membership voted to require members in public practice to participate in peer review. If a firm is a member of the AICPA Division of CPA Firms, either the SEC Practice Section (SECPS) or Private Companies Practice Section (PCPS), it undergoes a peer review. If the firm is not a member of the Division of CPA Firms, but its practice involves reporting on financial statements, it must enroll in either the Division of CPA Firms or the quality review program. For audits of governments, the General Accounting Office (GAO) established a mandatory triennial peer review requirement effective January 1989 (GAO 1986).

Similar to Francis et al. (1990), voluntary participation in the peer review program was in effect during the time period covered by this study. In a peer review, the review team performs a study and evaluation of the quality control system the reviewed firm had in effect during the period under review and then reviews selected accounting and auditing engagements to test the application of that system.

The purpose of the study and evaluation phase of the peer review is to determine whether the reviewed firm has designed quality control policies and procedures to meet the objectives specified by the AICPA in *Statement on Quality Control Standards No. 1*, "System of Quality Control for a CPA Firm," which prescribes the following nine elements of quality control: independence, assigning personnel to engagements, consultation, supervision, hiring, professional development, advancement, acceptance and continuance of clients, and inspection.

Following the review, the peer review team provides the firm with a report on its findings. Because the peer review program is intended to be positive, educational, and remedial, the constructive comments can result in improvement of the firm's practice. If the peer review team finds deficiencies in a firm's practices, remedial actions such as additional continuing professional education for firm personnel, control system changes, and special or accelerated reviews may be recommended.

As noted by Wallace and Wallace (1990), substantial practitioner literature is available describing peer review and its historical developments and the interaction between professionals and regulators. They examined specific letters of comment from the peer review files from 1980-1986. Bremser and Gramling (1988) provided evidence that SECPS member firms have improved quality control as a result of the peer review program. File, Ward, and Gray (1992) indicated that peer review may contribute toward enhancing the audit firm's credibility.

AUDIT ECONOMIC IMPLICATIONS OF PEER REVIEW

Extensive research has focused on the pricing of audit services, usually tested with log of audit fees as the dependent variable in a regression model. One line of research contends that this market induces differential pricing, because clients can choose the level of audit quality desired. Empirical studies generally confirm that Big Eight (now Big Six) firms charge higher fees, usually interpreted as a price premium paid for higher perceived audit quality (DeAngelo 1981; Francis 1984; Raman and Wilson 1992; Turpen 1990). Palmrose (1986) found that Big Eight firms expend more effort as measured by audit hours on their engagements. Deis and Giroux (1992) found that audit hours are significantly related to audit quality.

DeAngelo (1981) hypothesized that audit firms low-ball; that is, bid low (below first year audit costs) to obtain new clients. If the auditor retains the client, quasi-rents can be earned on subsequent audits. Quasi-rents represent a stream of future fees in excess of audit costs that arise because audit costs decline after the initial year when various startup costs and learning take place. DeAngelo (1981) believed that auditors will maintain high quality on the initial audit as an "investment" for future quasi-rents. Empirical studies supported lower audit fees on initial audits (Ettredge and Greenberg 1990; Roberts, Glezen, and Jones 1990; Turpen 1990). However, it is difficult to determine audit economic relationships unless audit quality can be measured.

Francis et al. (1990) asked if peer reviewed firms were perceived as quality differentiated auditors. If so, fee premia should be charged. A fee premium, if identified, would act as an economic incentive, suggesting a mandatory peer review program is not necessary, according to Francis et al. (1990). Without a premium no economic incentive exists and a mandatory program would be required (assuming that peer review is a needed regulatory program). Their testing, based on a multivariate audit fee model, indicated no significant fee differences for peer reviewed firms (using AICPA Division of CPA Firms membership as a surrogate for peer review).

The peer review is not a direct measure of audit quality, however. Much like a Big Eight dummy variable, it is assumed that a peer reviewed firm has higher audit quality than a non peer reviewed firm. The measure of audit quality is a separate construct. This study extends Francis et al. (1990) to the governmental sector where an independent measure of audit quality exists. Two hypotheses are examined.

Hypothesis 1. Peer reviewed audit firms are quality differentiated.

Hypothesis 2. Peer reviewed audit firms command audit fee premia.

The hypotheses are analyzed using both univariate and multivariate models.

Peer review is a professional self-regulatory program. It is effective if the perceived benefit is a higher quality audit. The Single Audit Act of 1984 provided a mechanism for evaluating audit quality through required working paper reviews of auditors of state and local governments by federal and state regulatory agencies.[1]

Deis and Giroux (1992) developed an audit quality metric of Texas school district audits based on information contained in detailed questionnaires used in reviews of auditor working papers. Deis and Giroux (1992) based their metric on findings issued by the Texas Education Agency's Director of Audits to the audit firm. The comments refer to various deficiencies or omissions discovered during the working paper review. For analysis, the findings were grouped into nineteen categories which were then ranked by the Director of Audits according to perceived importance in the decision to declare audit work substandard and, accordingly, necessary to refer the auditor to the Texas State Board of Public Accountancy for possible disciplinary action. As a result, Deis and Giroux (1992) were able to generate a metric that captures both the presence and, qualitatively, the relative importance of various audit deficiencies. For example, failure to use an audit program was the most important deficiency, failure to issue an engagement letter was the fourteenth most important, and the omission of a management representation letter was considered the sixth most important deficiency (see Deis and Giroux 1992, 469). A similarly designed metric is used in this study to proxy for audit quality. It is expected that peer reviewed audit firms have significantly higher audit quality scores than similar nonpeer reviewed firms.[2]

Following Francis et al. (1990) it is expected that peer reviewed firms will charge higher audit fees as quality differentiated auditors. Several studies using Big Eight firms as a "brand name" surrogate for quality found higher fees associated with Big Eight auditors (e.g., DeAngelo 1981; Baber, Brooks, and Ricks 1987; Turpen 1990). Francis et al. (1990) proposed that voluntary membership in the Division is potentially another form of "brand name" available to smaller (e.g., non-Big Eight) auditors. Big Eight firms were excluded from the study by Francis et al. (1990) to mitigate their confounding "brand name" effect. The Big Eight "brand name" effect also has been documented in the public sector market (Raman and Wilson 1992); hence, our sample is also restricted to the small audit firm segment of suppliers.[3]

Francis et al. (1990) found no fee premia for AICPA Division for CPA Firms, used as a surrogate for peer review. Given the absence of fee premia, they concluded that there is an economic disincentive to join the Division. The public sector audit market, however, differs from the private sector in environment and contracting procedures (Rubin 1988). Voluntary participation in the AICPA's peer review program offers auditors of public sector agencies a mechanism to signal quality in a competitive market that includes low quality suppliers.[4]

SAMPLE

This study considers audits of Texas independent school districts (ISDs), based on ISDs subject to quality control or working paper reviews (QCRs) conducted by the Texas Education Agency (TEA) for fiscal years 1985 to 1988. The TEA analyzed working papers at the CPAs' offices to determine if sufficient evidence was gathered by the auditor to support representations made in the auditor reports. After the on-site review, the TEA sends a findings letter to the CPA firms. Few QCRs have "clean" reviews (no exceptions). In most cases the letter makes a list of recommendations and suggestions for improving the audit. Significant deficiencies may result in a referral to the Texas State Board of Public Accountancy. A total of 308 QCRs were conducted by the TEA during this period. The final sample for this project was 232 audits, which excluded the early (1983 and 1984 fiscal years) QCRs which were conducted before the program was fully developed, Big eight and national audit firms because there were only ten, and eight QCRs when the same audit firm was reviewed more than once in the same year.[5]

Most of the data analyzed were provided from TEA sources, including QCR findings letters by special permission of the TEA. Audit fees and audit hours, auditor tenure, and AICPA Peer Review membership were gathered from the QCR files. Financial and other ISD data were available from TEA computer files. The data are described in the next section.

DESCRIPTIVE ANALYSIS OF AUDITOR CHARACTERISTICS

Table 1 summarizes key descriptive statistics of the sample. Five categories of data are presented. Many variables, such as auditor tenure and auditor change, are common to the audit economics area and are included primarily for control purposes.

The first category includes fee and hour information. As expected, fees and hours vary. Of particular interest is average fee per hour. Most of the audit firms are relatively small, often a single practitioner. Consequently, different billing rates (by staff levels) are less of a problem. The average fee per hour is $38.90.[6] Two possible explanations are suggested for the relatively low fees. First, the ISD audit seems to be highly competitive in Texas. Second, these audits are conducted from September to December, generally a "slow" audit period for small audit practices.

Table 1. Sample Distribution Characteristics/Frequencies
(n = 232)

	Means/ Frequencies	Standard Deviations/ Percentage	Minimum	Maximum
Fees/Hours				
Audit Fee (FEE)	$10,416	8,915	800	50,000
Audit Hours (HOURS)	294.4	260.9	38	1,994
Fees/Hour (FEEHOUR)	$ 38.90	18.5	2.4	172.6
Audit Quality (QCR)				
Quality Score (SCORE)	– 36.9	28.7	–157	0
Referred Audits (REFER)	32	14%		
Auditor Tenure				
Tenure in Years (TENURE)	10.1	8.8	1	40
Auditor Change (YEAR1)	22	10%		
Second Year (YEAR2)	20	9%		
Other Auditor Characteristics				
Peer Review (PEER)	48	21%		
Member of AICPA (MEMBER)	95	41%		
Percent of 120 Days to Submit Report to TEA (TIME)	.63	.23	5	1.61
Late Reports (LATE)	7	3%		
Number of ISD Clients (CLIENTS)	3.3	3.9	1	30
Client Size				
Average Daily Attendance (ADA)	3,615	5,867	36	44,776

Two variables are included to measure minimum audit quality based on TEA QCR results. As in Deis and Giroux (1992), SCORE is a qualitative measure based on QCR findings' letters submitted to the auditors by the TEA. A letter indicating no exceptions receives a score of 0. Negative scores relate to the number of quality deficiencies multiplied by the severity of the deficiencies. The Director of Audits of the TEA ranked the importance of each deficiency, from 1 (least important) to 19 (most important) relative to the referral decision. A metric of audit quality was constructed by multiplying each deficiency category by its (negative) rank, from ⁻1 to ⁻19. The weights are summed, resulting in the metric SCORE. The average score of ⁻37 suggests that substantial deficiencies were the norm rather than the exception. REFER is a dummy variable where 1 represents an auditor referred to the Texas State Board for possible disciplinary action. Thirty-two auditors were referred over the four-year QCR period because of excessive deficiencies.

Auditor tenure averaged over 10 years. Only 22 ISDs had an auditor change the year of the QCR, while 49 ISDs had the same auditor at least 20 years. Thus, auditor turnover was uncommon. This differs from Francis et al. (1990) where approximately one-third of the sample changed auditors in a 3-year period. (The TEA began to issue audit procurement regulations after the period of study.)

Six additional auditor characteristics are analyzed. Ninety-five firms were members of the AICPA, 48 of which were peer reviewed by the AICPA. None of the peer reviewed firms was referred to the State Board. Most audit firms finished the audit early and the report was submitted well ahead of the December 31 deadline. Only seven reports were submitted late. Dwyer and Wilson (1989), Deis and Giroux (1992), and Rubin (1992) found consistent evidence that report timeliness reflects audit quality, with late reports associated with lower quality. The typical firm had 3.3 ISD clients; 115 firms had only one, while 19 firms had at least ten. Finally, only 31 firms had more than a single office. Auditors for most Texas ISDs are relatively small firms.

Average daily attendance (ADA) is used as a measure of school district size. Knapp (1991) suggested that client size is negatively associated with audit quality because larger clients can resolve more conflicts in their favor. Deis and Giroux (1992) concluded that the potential for "complacency" by the auditor and the power exercisable by clients may increase the potential for lower quality audits. Copley and Doucet (1993) also observed a negative, but not significant, relationship between client size and a dichotomous coding of audit quality. Copley et al. (1994), however, find a significant positive relationship between the client size (i.e., total budgeted revenues or level of federal funding) and a dichotomous variable indicating acceptability of the audit report. Data specificity most likely accounts for these conflicting results. The Copley et al. (1994) study contains higher profile audits of federal agencies and funds (e.g., HUD audits and higher education audits), whereas Deis and Giroux (1992) contains arguably lower profile audits of local independent school districts. Through repeated announcements of substandard audit quality, one might expect more uniform audit quality at some point (but this is not expected from the 1985-1988 audits used by the two studies).

Average size is 3,600 students, with a range of 36-45,000. Client size explains most of the audit fee variance in typical

Table 2. Sample Partitioning by Peer Review and AICPA Membership
Mean Comparisons Using Selected Ratios: t-tests (Wilcoxon)

	Peer Review		t-test/	AICPA Member		t-text/
	No	Yes	Wilcoxon	No	Yes	Wilcoxon
FEE	$9,587	$13,595	−2.82 (3.41)**	$9,672	$11,489	−1.46 (1.34)
HOURS	281	347	−1.58 (2.42)***	272	327	−1.47 (1.47)
FEEHOUR	$38	$41	−0.94 (1.41)	$38	$40	−0.57 (0.27)
SCORE	−41	−23	−6.15 (4.04)*	−36	−38	0.05 (−0.39)
TENURE	10	11	−0.93 (1.10)	11	9	0.68 (−0.84)
ADA	3,350	4,632	−1.36 (2.13)***	3239	4158	1.24 (1.38)
n	184	48		137	95	

Notes: * significant at .0001
** significant at .01
*** significant at .1

fee studies. This paper, however, also considers fee per audit hour
and quality. No prediction is made for this relationship. Economies
of scale may exist, but it is not known if the auditor would pass along
the cost savings to the client. The relationship to quality is unknown.

The descriptive analysis is extended by partitioning the sample by
peer review and AICPA membership in the form of dummy variables
and by comparing certain ratios. Mean differences are presented in
Table 2. Forty-eight firms volunteered for peer reviews (i.e., 48 of
the 95 AICPA members). These firms charged slightly more per hour
and they audited larger school districts, while size and quality scores
were significantly different based on *t*-tests or Wilcoxon 2-sample
tests. Interestingly, no difference was noted between firms with or
without membership in the AICPA.

MULTIVARIATE ANALYSIS

Regression is used to analyze (1) log of audit fees (LOGFEE), (2)
log of audit hours (LOGHOUR), and (3) audit fees per hour

(FEEHOUR). Audit quality as measured by QCR results (SCORE) is analyzed using analysis of variance (ANOVA). LOGFEE is the log of total audit fees for the ISD audit, the traditional fee model. LOGHOUR is the log of total audit hours worked on the ISD engagement, also used by Palmrose (1989). A separate FEEHOUR model was added to be comparable with the univariate analysis. SCORE, unlike the measure used by Deis and Giroux (1992), was constructed such that higher values of SCORE indicate higher audit quality.[7]

Fee Models

The LOGFEE, LOGHOUR, and FEEHOUR models are:

$$\text{LOGFEE, LOGHOUR, FEEHOUR} =$$
$$\text{YR1} + \text{YR2} + \text{REFER} + \text{PEER} + \text{CLIENTS} + \text{SIZE} \qquad (1)$$

YR1 and YR2 are low-balling-related variables with YR1 the year of the auditor change and YR2 the following year. Because the auditor change market is considered competitive, LOGFEE and FEEHOUR should be significantly lower than in later years. DeAngelo (1981) predicted that fees should be lower and audit costs higher in year one. Deis and Giroux (1993) found both to be true in separate fee and hour regression runs of ISD audits. DeAngelo (1981) expected "normal fees" by year two, in the sense that auditors will be earning anticipated quasi-rents beginning in year 2. However, some empirical studies found higher fees through year 2 (Turpen 1990) or year 3 (Simon and Francis 1988). It is expected that the coefficient will be negative for YR1 and nonpositive for YR2.

DeAngelo's (1981) framework anticipates that audit firms will maintain a relatively high level of audit quality, at least in the initial years, to ensure that the "investment" (i.e., price discounts on initial audits) in the client is protected. *SAS 55* and other auditing literature speaks about the importance of understanding the client in guiding the audit approach. Hence, first-year audits necessarily involve more formal documentation due to the learning phase of the engagement relative to later years' work. Accordingly, more audit hours are expected in first-and possibly second-year audits. It is expected, therefore, that the coefficient for YR1 and YR2 will be positive. This expectation is congruent to Palmrose (1989) where she determined that price discounts associated with fixed fee audit contracts were not impounded into audit hours.

Referred audits have quality so low that the TEA referred the auditors to the Texas State Board for remedial or punitive action. It is expected that these audits have lower fees (LOGFEE), but also lower hours (LOGHOUR). Therefore, no prediction is made for the FEEHOUR model, while a negative coefficient is expected for the other two models.

Auditors volunteering for a peer review presumably are signalling to present and potential clients that they meet all minimum audit requirements for a broad range of industry specialties. This signaling should allow them to charge a higher fee, but greater audit effort would be expected. Thus, a positive prediction is made for PEER in all three models.

Number of independent school district clients indicates an industry specialization in ISD audits, with associated expertise and potential economies of scale. Both higher fees (LOGFEE) and lower hours (LOGHOUR) are expected.

SIZE, as measured by log of ADA, is a control variable. The audit market may vary by client size. Obviously, total audit fees are larger for bigger clients, a positive sign for the LOGFEE model. However, this should be offset by more audit hours (LOGHOUR). If economies of scale exist, fees per hour are negatively related to size assuming that the auditors pass on the cost savings to the clients. No direction is predicted for FEEHOUR.

Results for the LOGFEE, LOGHOUR, and FEEHOUR models are presented in Table 3. Several diagnostic tests were run on both models to ensure that regression assumptions were met. No severe statistical violations were detected.[8]

The LOGFEE model has an adjusted R^2 of 60%, significant at .0001 and similar to other studies. Four of the six independent variables were significant in the expected direction. YR2 was not significant, indicating that low-balling exists only for the year of the auditor change in our sample. CLIENTS was insignificant, suggesting no fee premium for industry specialization.

The LOGHOUR model has an adjusted R^2 of 43%, significant at .0001. All six explanatory variables had the expected signs and four were significant. YR2 was not significant, indicating that most of the

Table 3. Regression Results for Fee and Hour Models
(n = 232)

	LOGFEE		LOGHOUR		FEEHOUR
	Sign	Coefficient (t-value)	Sign	Coefficient (t-value)	Sign	Coefficient (t-value)
Intercept		6.11		3.10		15.34
YR1	−	−.14 (−1.35)***	+	0.16 (1.35)***	−	−11.03 (2.80)***
YR2	−	.11 (.17)	+	0.06 (0.49)	?	−3.15 (−.77)
REFER	?	−.25 (−2.76)**	−	−0.42 (3.95)*	?	9.84 (2.87)**
CLIENTS	+	−.01 (−1.23)	−	−0.02 (1.94)**	+	0.19 (.65)
SIZE	+	.40 (17.87)*	+	0.33 (12.51)*	?	3.05 (3.59)**
PEER	+	.17 (2.19)***	+	0.08 (0.89)	+	2.16 (.74)
R^2		.605		.427		.113
F Statistic		59.9*		29.64*		5.91*

Notes: *significant at .0001
**significant at .01
***significant at .1

client learning takes place in the first year. PEER was insignificant, indicating that additional audit hours do not result from being a member of a PEER review program. Of course, this result may reflect firms that more efficiently conduct audits due, in part perhaps, to PEER review participation.

The FEEHOUR model has an adjusted R^2 of 11%, significant at .0001, but with considerably lower explanatory power than either the fee model or hour model (which is explained largely by client size). Only three variables are significant: SIZE, YR1, and REFER. SIZE has a positive coefficient, indicating that economies of scale do not exist (or if they do the cost savings are not passed on to the client in the form of lower fees per hour). Instead, larger districts pay relatively more per audit hour. Clients whose auditors are referred have the dubious distinction of paying more per audit hour (basically because relatively fewer hours were used during the audit). Fees per hour were lower in the year of an auditor change (but not the following year), consistent with low-balling. Clients did not pay a premium price for either auditors subject to a peer review or for ISD specialization.

Audit Quality Model

Following Knapp (1991) an analysis of variance (ANOVA) model is used to analyze audit quality. DeAngelo (1981) defined audit quality as the probability the auditor will discover and report an accounting breach. Discovery depends on the auditor's technical capabilities and audit effort. SCORE is a surrogate of minimum audit quality based on an external measure of quality, a working paper review by a TEA audit review team. SCORE is the dependent variable of the audit quality model.

The quality model is:

$$SCORE = PEER + TENCAT + CLICAT + LATE \qquad (2)$$

A peer review (PEER) is somewhat similar to a QCR. An audit firm volunteering for a peer review should be more successful during a QCR.

A tenure categorical variable (TENCAT) is used, which is similar to Knapp (1991). Auditor tenure is analyzed by category. Several authorities suggest that new auditors (especially first-year auditors) are subject to low quality, based on a high frequency of substandard audits in the early years (e.g., AICPA 1986 and a "learning curve" effect that places a new auditor at a disadvantage (e.g., DeAngelo 1981). On the other hand, a long association between auditor and client can limit audit quality because of complacency and less rigorous audit procedures (Shockley 1981). As auditor tenure increases, auditor independence may decline and the auditor may be less likely to report irregularities. Knapp (1991) suggests that quality declines after 20 years. Four tenure categories are used:

1. first-year audit;
2. second-, third-, and fourth-year audits;
3. audits of 5 to less than 20 years; and
4. audits of 20 years, and over.

A firm specializing in ISD audits should achieve a higher quality score. Number of clients is used as a surrogate for specialization, measured as a three-level categorical variable: (1) audit firms with a single ISD client (no specialization), (2) firms with 2-4 ISD clients, and (3) firms with five or more ISD clients (high specialization).

Complete audited financial statements and required reports must be submitted to the TEA by the end of the calendar year, 120 days after the end of the fiscal year. A dummy variable (LATE) is used where 1 represents statements submitted after the deadline. It is expected that statements submitted late were subject to audit problems, likely to result in low QCR scores.

Multiple comparison tests (Duncan's Multiple Range Test) are run for TENCAT and CLICAT to determine significant differences by category. This is particularly important for TENCAT because a negative relationship is predicted for both first-year audits and audits of 20 years or more.

ANOVA results are presented in Table 4. The R^2 is 14%, significant at .001. All independent variables are significant. The increase in quality associated with PEER indicates that a voluntary peer review is an effective signal of quality. Auditors with more ISD clients have greater quality, suggesting the importance of specialization. Finally, reports that are submitted late are associated with lower quality.[9]

Duncan test results are presented in Table 5. The four tenure categories of TENCAT are analyzed in Panel A. First year audits have the highest quality score (an unexpected result), which is significantly higher than category 4 (audits of 20 or more years).

Table 4. ANOVA Results for SCORE

Description	Coefficient	Mean Square	F-Value
Intercept	−3.457		
PEER	0.439	12,041	16.47*
TENCAT	−.109	2,437	3.33***
CLICAT	.185	2,617	3.58***
LATE	−0.521	2,492	3.41***

F-value = 5.29, significant at .0001.
R^2 = .142; n = 232

Notes: * significant at .0001
** significant at .01
*** significant at .1

Table 5. Duncan Multiple Range Test for TENCAT and CLICAT
Dependent Variable = SCORE
(significant difference at .05)

	Mean	Grouping		n
Panel A Tenure Category				
1. First Year	−27.0	A		22
2. 2-4 years	−37.7	A	B	64
3. 5-19 years	−34.5	A	B	97
4. 20 + years	−45.3		B	49
Panel B Number of ISD Clients				
1. 1 client	−39.5		B	115
2. 2-4 clients	−40.0		B	65
3. 5 or more clients	−27.3	A		52

This suggests that the new auditor makes a greater audit effort, consistent with DeAngelo's (1981) observation that the initial audit is an "investment" for future quasi-rents. Quality declines by tenure category, consistent with increasing complacency and reduced auditor independence.[10] CLICAT is analyzed in Panel B. Auditors specializing in ISD audits (category three or auditors with five or more ISD clients) have significantly higher quality scores, as expected. There is no significant difference between categories one and two.

CONCLUSIONS

This paper extends the audit economics literature by modeling audit fees and audit quality with particular emphasis on peer review. The TEA provided a rich database on Texas ISDs that allowed extensive analysis of relevant data. Fee analysis indicated significance based on auditor changes and tenure, client size, and number of ISD clients. Results support low-balling and provide further analysis on the relationship of audit fees to audit hours. Also, auditors referred to the State Board charged a higher fee per hour for these audits. One explanation consistent with this result is that the auditor bids at or near market price but shirks in effort supplied, thereby inflating the fee per hour measure.

Audit quality was strongly associated with auditor characteristics tested, including auditor tenure, peer review participation, number of ISD clients, and late reports. Peer reviewed auditors have higher quality scores as expected. Audits submitted after the year-end due date had lower quality. Tenure and auditor specialization were further analyzed using multiple comparison tests. Tenure results were much simpler than anticipated: quality declined as tenure increased. First-year audits had the highest quality scores, audit tenure of 20 or more years the lowest. Whether because of complacency, lack of independence, or other factors, this is a disturbing relationship. Specialization as measured by auditors with five or more ISD clients also resulted in significantly higher quality.

Peer review and other auditor characteristics are important factors when analyzing Audit Economics relationships and should be incorporated in future research. When considering the multivariate models, results for auditor tenure and ISD clients are of interest. The client results suggest that auditors with ISD specialization enhance the quality of the audit. The tenure results suggest that ISDs should consider a policy of replacing auditors periodically. First-year audits have the highest quality, at a significantly lower audit fee. Quality declines with auditor tenure.

Particularly important are the peer review-related findings. Peer reviewed audit firms charge significantly higher audit fees; however, there are no fee differences on a per hour basis. This suggests that higher fees are explained by greater audit effort. Peer review is significantly related to quality scores, indicating higher quality audits. Also, no peer reviewed firm was referred to the Texas State Board. In summary, peer review was associated with higher quality and higher fees, but the higher fees can be explained by greater audit effort. These results can be used to support the usefulness of mandatory peer review in the public sector.

NOTES

1. In the Single Audit Act of 1984, cognizant agencies were established and charged with the responsibility to implement the requirements of the act. Pertaining to the auditor, some of the more important cognizant agency responsibilities include: (1) providing technical advice, (2) conducting desk reviews of audit reports, (3) conducting working paper reviews of audit organizations, (4) resolving deficiencies noted in desk and working paper reviews, and (5) processing finalized audit reports.

2. Deis and Giroux (1992) provide evidence consistent with this interpretation. Their model regressing a score for audit quality on a set of variables indicated better quality audits associated with CPA firms voluntarily participating in the AICPA's peer review program. Deis and Giroux did not investigate the fee structure of the audit.

3.   Of particular relevance to this study is Raman and Wilson's (1992, 292) claim that they found their sample of city auditees to be "cognizant of product differentiation in auditing" with a brand name fee premium observed in the small auditee market.

4.   Reports concerning nonfederal audit quality issued by the President's Council on Integrity and efficiency (PCIE) and the Texas State Board of Public Accountancy both attest to the alarmingly high rate of "problem" audits. A January 26, 1994 memo from the PCIE's Standards Subcommittee reports 47% of reports subjected to QCRs were found to require major changes or were significantly inadequate. Moreover, the Committee was troubled by a "decrease in quality" from previous examinations. Generally, one-quarter to one-third of audits reviewed over the last ten years have been deemed substandard or inadequate in some phase of the audit.

5.   The QCR engagement was occasionally expanded to include additional audit reports at the discretion of the review team, when extremely substandard audits were encountered. In such cases, only the initial QCR was included in the study. Moreover, the follow-up QCRs were not always complete as the team may have been investigating the pervasiveness of a particular audit deficiency.

6.   This is comparable to an average fee per hour of $37.45 derived by TEA in a study of 1989 audit fees (for school districts similar in size to those included in this study).

7.   Deis and Giroux (1992) constructed a variable (QUALITY) so that higher values of QUALITY reflected lower overall audit quality.

8.   Variance inflation factors (VIFs) were below two for all variables in both models, indicating no severe multicollinearity. Univariate tests of the residuals indicate mean zero and near-normal distributions. Both the Parks Test and the White's Test were used to test for heteroscedascity and no violations were found. Additionally, the models were conducted omitting, successively and in combination, observation with the five smallest FEEHOUR values (i.e., $2.41 to $12.20) without any differences being detected in the regression results. Hence, model results reported are for the full 232 observation sample.

9.   The ANOVA model was initially run with all two-way interactions. None of these was significant and, therefore, not presented in the final model.

10.   An anonymous reviewer points out, however, that first-year audits, which generally accumulate more documentation than subsequent year audits, may create a systematic bias if the relative quantity of information collected influences the QCR review team. This is an important point to consider when constructing a metric based on QCR results. The metric used in this study, SCORE, isolates aspects important in all audits and, hence, should avoid the bias the reviewer warns of.

REFERENCES

American Institute of Certified Public Accountants. 1986. *Restructuring Professional Standards to Achieve Professional Excellence in a Changing Environment*. New York: AICPA.

Baber, W., E. Brooks, and W. Ricks. 1987. An empirical investigation of the market for audit services in the public sector. *Journal of Accounting Research* (Autumn): 293-305.

Bremser, W., and L. Gramling. 1988. CPA firm peer reviews: Do they improve quality? *The CPA Journal* (May): 75-77.

Copley, P., and M. Doucet. 1983. The impact of competition on the quality of governmental audits. *Auditing: A Journal of Practice and Theory* (Spring): 88-98.

Copley, P., M. Doucet, and K. Gaver. 1994. A simultaneous equations analysis of quality control review outcomes and engagement fees for audits of recipients of federal financial assistance. *The Accounting Review* (January): 244-256.

DeAngelo, L. 1981. Auditor size and audit quality. *Journal of Accounting and Economics* (December): 183-199.

Deis, D., and G. Giroux. 1992. Determinants of audit quality in the public sector *The Accounting Review* (July): 462-479.

———. 1993. Public sector evidence of product differentiation effects on audit price and audit effort. Working paper, Department of Economics, Texas A&M University, College Station.

Dwyer, P., and E. Wilson. 1989. An empirical investigation of factors affecting the timeliness of reporting by municipalities. *Journal of Accounting and Public Policy* (Spring): 29-55.

Ettredge, M., and R. Greenberg. 1990. Determinants of fee cutting on initial audit engagements. *Journal of Accounting Research* (August): 198-210.

File, R., B. Ward, and C. Gray. 1992. Peer review as a market signal: Effective self-regulation? In *Research in Accounting Regulation*, Vol. 6, ed. G.J. Previts, 179-193. Greenwich, CT: JAI Press.

Francis, J. 1984. The effect of audit firm size on audit prices: A study of the Australian market. *Journal of Accounting and Economics* (August): 133-151.

Francis, J., W. Andrews, and D. Simon. 1990. Voluntary peer reviews, audit quality, and proposals for mandatory peer reviews. *Journal of Accounting, Auditing & Finance* 5(3): 369-378.

General Accounting Office. 1986. *Audit Quality: Many Governments Do Not Comply With Professional Standards*. Washington, DC: GAO.

Granof, M. 1992. Privatization: The road to federal regulation of auditing. *Accounting Horizons* (December): 76-85.

Knapp, M. 1991. Factors that audit committee members use as surrogates for audit quality. *Auditing: A Journal of Practice and Theory* (Spring): 35-52.

Palmrose, Z. 1986. Audit fees and auditor size: Further evidence. *Journal of Accounting Research* (Spring): 97-110.

———. 1989. The relation of audit contract type to audit fees and hours. *The Accounting Review* (July): 488-499.

Raman, K. and E. Wilson. 1992. An empirical investigation of the market for "single audit" services. *Journal of Accounting and Public Policy* (Winter): 271-295.

Roberts, R., G. Glezen, and T. Jones. 1990. Determinants of auditor change in the public sector. *Journal of Accounting Research* (Spring): 220-228.

Rubin, M. 1988. Municipal audit fee determinants. *The Accounting Review* (April): 219-236.

_____. 1992. Recent public choice research relevant to government accounting and auditing. In *Research in Governmental and Nonprofit Accounting*, Vol. 3, Part A, ed. J. Chan, 129-145. Greenwich, CT: JAI Press.

Shockley, R. 1981. Perceptions of auditor independence: An empirical analysis. *The Accounting Review* (October): 785-800.

Simon, D., and J. Francis. 1988. The effects of auditor change on audit fees: Tests of price cutting and price recovery. *The Accounting Review* (April): 255-269.

Turpen, R. 1990. Differential pricing on auditors' initial engagements: Further evidence. *Auditing: A Journal of Practice and Theory* (Spring): 60-76.

Wallace, W. 1989. Historical views of the SEC's reports to Congress. *Accounting Horizons* (December): 24-39.

Wallace, W., and J. Wallace. 1990. Learning from peer review comments. *The CPA Journal* (May): 48-53.

REDUCING THE INCIDENCE OF
FRAUDULENT FINANCIAL REPORTING:

EVALUATING THE TREADWAY COMMISSION
RECOMMENDATIONS AND
POTENTIAL LEGISLATION

Jerry R. Strawser, John O'Shaughnessy, and
Philip H. Siegel

ABSTRACT

The Treadway Commission provided recommendations to public
companies, independent public accountants, and oversight bodies to
enhance the reliability of the financial reporting process. The ultimate
goal of the Treadway Commission's report (National Commission on
Fraudulent Financial Reporting 1987) is to reduce the incidence of
fraudulent financial reporting in the United States. This paper presents
the results of a survey of internal audit managers regarding the
implementation of measures to reduce the incidence of fraudulent

Research in Accounting Regulation, Volume 9, pages 83-104.
ISBN: 1-55938-883-8

financial reporting and their perceptions of the effectiveness of potential legislation in reducing the incidence of fraudulent financial reporting. The measures and legislation examined herein include recommendations of the Treadway Commission as well as other items determined through a pretest. Details of this test are provided in the text. The results of this survey indicate that, in general, public companies are in compliance with the recommendations of the Treadway Commission; however, some exceptions (particularly the existence of a quality assurance program and the use of the IIA's Professional Standards [IIA 1978] as evaluation criteria) do continue to exist. Also, several proposed forms of legislation appear to be perceived by internal audit managers as being effective in preventing and detecting fraudulent financial reporting. The findings of this survey have implications for public companies, their internal audit functions, and oversight bodies (such as the Securities and Exchange Commission).

In 1985, the National Commission on Fraudulent Financial Reporting (Treadway Commission) undertook a comprehensive study of the financial reporting system of public companies in the United States. In performing this investigation, the Treadway Commission identified three major groups of participants in the financial reporting process: public companies, independent public accountants, and oversight bodies (e.g., the Securities and Exchange Commission). Included in the final report of the Treadway Commission (National Commission on Fraudulent Financial Reporting [NCFFR] 1987) are recommendations for each of these groups to improve the practice of financial reporting in the United States and reduce the incidence of fraudulent financial reporting.[1] This report also includes a number of suggestions that may enable companies to achieve these recommendations and ultimately, reduce the incidence of fraudulent financial reporting.

In discussing its recommendations for public companies, the Treadway Commission indicates that fraudulent financial reporting should be addressed by both upper management as well as functions within the company. Included in this latter category is the company's internal audit function. Because of the role of the internal audit function in monitoring the reliability and integrity of financial information, it appears that internal auditors play a significant role in reducing the incidence of fraudulent financial reporting. Consistent with this role, the Treadway Commission's report provides extensive recommendations relating to the role of the internal audit function in the financial reporting process.

This paper presents the results of a survey of internal audit managers regarding the implementation of measures to reduce the incidence of fraudulent financial reporting and their perceptions of the effectiveness of potential legislation in reducing the incidence of fraudulent financial reporting. The measures and legislation examined in this survey include recommendations of the Treadway Commission as well as other items determined through a pretest. A preliminary questionnaire was developed which incorporated the Treadway Commission's internal audit recommendations and implementation suggestions. The questionnaire was pretested on a cross-section of internal audit managers, CPAs, and accounting educators. The instrument had three sections: a detailed demographic section, questions using a three-point Likert scale to determine the extent of implementation of the Treadway Commission's recommendations, and several open-ended questions. The purpose of this survey is twofold. First, information on the extent to which measures have been implemented by a sample of public companies provides some indication as to the effect of the Treadway Commission's report on the practice of internal auditing. Second, identifying the perceived effectiveness of ten potential regulatory actions on the occurrence of fraudulent financial reporting may allow oversight bodies to enhance the integrity of the financial reporting process.

BACKGROUND

In setting forth its recommendations to public companies, the Treadway Commission identified two levels of individuals: upper management (Chief Executive Officer, Chief Financial Officer, and the board of directors) and functions within the company that are important to the integrity of financial reporting (accounting function, internal audit function, and audit committee of the board of directors). While both the accounting function and audit committee are important in reducing the incidence of fraudulent financial reporting, the role played by the internal audit function in monitoring the reliability of financial information and operations provides this function with a unique opportunity to detect instances of fraudulent financial reporting.[2] The Treadway Commission provided four recommendations for public companies related to their internal audit functions. These include (1) establishing effective internal audit functions, (2) ensuring the objectivity of their internal audit functions, (3) considering the implications of nonfinancial internal audit findings, and (4) ensuring that the internal auditors' involvement in the financial reporting process is coordinated with that of the independent auditor. These recommendations are briefly discussed in the following paragraphs.

An effective internal audit function is evidenced by the strong support of both upper management and the audit committee. To achieve this degree of support, the Treadway Commission recommends that public companies establish a written charter for the internal auditing department formally outlining the department's role and authority within the company. In addition, the Treadway Commission notes that companies can increase the effectiveness of their internal audit functions by offering continuing professional education and providing attractive career paths to internal audit staff. Finally, the Treadway Commission recommends that public companies adopt the *Standards for the Professional Practice of Internal Auditing*[3] (Institute of Internal Auditors 1978) (hereafter, IIA Standards) and undergo periodic peer reviews of the work of the internal audit function. These peer reviews (known as quality assurance reviews) should allow public companies to identify departures from the IIA Standards and enhance the effectiveness of the work performed by the internal audit function.

The second recommendation of the Treadway Commission relates to the objectivity of the internal audit function. This objectivity is related to both the organizational position of the function and the position of the chief internal auditor within the organization. In order to be objective, the internal audit function should be free to examine any area of the company's operations and report the findings of their examination to the appropriate parties. The Treadway Commission recommended that the chief internal auditor should: (1) report administratively to an officer not directly involved in the preparation of the company's financial statements (such as the CEO), (2) have direct and unrestricted access to the CEO and audit committee, and (3) regularly attend all audit committee meetings and report to the audit committee at regular intervals. Because of the importance of the internal audit function to the activities of the audit committee, a final recommendation is that the audit committee should review the appointment and dismissal of the chief internal auditor.

Two final recommendations relate to considering the implications of nonfinancial audit findings and coordinating the internal auditor's involvement in the independent audit examination. Internal auditors should consider the results of nonfinancial audits (e.g., operational audits, acquisition audits, and special investigations) in evaluating the existence of fraudulent financial reporting. Also, given their knowledge of the company and its internal control structure, the internal auditor's involvement in the external audit may allow further instances of fraudulent financial reporting to be detected. Consistent with this recommendation, the IIA has issued *Statement on Internal Auditing Standards No. 5* (IIA 1987) to provide guidance on coordinating the efforts of these two parties.

This study provides an indication of the extent to which Treadway Commission recommendations are implemented by a sample of public companies. We identify how the incidence of fraudulent financial reporting is affected by both (1) recommendations of the Treadway Commission and (2) other potential actions. Identifying specific recommendations and actions that effectively reduce the incidence of fraudulent financial reporting allows public companies and regulatory bodies to consider these recommendations and actions in establishing the roles and responsibilities of their internal audit functions.

THE SURVEY

The internal audit managers of the 100 largest publicly held companies located in the state of California were asked to complete a brief questionnaire relating to fraudulent financial reporting.[4] This response rate was calculated using the following ratio:

Response Rate   =   [(complete responses received)/(total sample size, including replacements) – (companies without internal audit departments)]

= $74/(100\text{-}9) = 81.3\%$

This questionnaire elicited three primary types of information that is analyzed in this study. First, internal audit managers were asked to indicate the extent to which various measures have been implemented by public companies and their internal audit functions to reduce the incidence of fraudulent financial reporting. These measures include recommendations of the Treadway Commission as well as other measures identified through the pretest. Internal audit manager responses as to the extent of implementation were elicited using a three-point scale: "not [implemented] at all," "[implemented] to some extent," and "[implemented] to a great extent."

The second type of information elicited in the questionnaire was the perceived effectiveness of ten potential forms of regulation or legislation in preventing the occurrence of fraudulent financial reporting for public companies. Measuring the perceived effectiveness of various forms of proposed legislation may allow companies to reduce the incidence of fraudulent financial reporting by identifying measures most effective in preventing or detecting instances of fraudulent financial reporting.

Finally, information was obtained about the internal audit manager and the sample company. Internal audit managers were asked to provide demographic information related to their years of experience and professional certification. While not directly elicited from internal audit managers, the size of the sample company was identified from publicly available sources. Total revenues of the firm were used to identify the 100 largest publicly held companies in California. The data were used to determine whether the implementation of measures to reduce the incidence of fraudulent financial reporting or the perceived effectiveness of various measures in this regard are affected by characteristics of the internal audit manager or sample company.

RESULTS

Implementation of Treadway Recommendations

Table 1 summarizes the internal audit managers' responses regarding the extent to which measures that may reduce the incidence of fraudulent financial reporting have been implemented by sample companies. The audit managers' responses are separately discussed by category in the following subsections.

Table 1. Implementation of Measures to Reduce the Incidence of Fraudulent Financial Reporting

	Not at All	To some Extent	To a Great Extent
Effectiveness of the Internal Audit Function			
1. The responsibility of the internal audit function has been set forth in writing*	3 (4.1)	22 (29.7)	49 (66.2)
2. The company financially supports continuing education of internal audit staff*	3 (4.1)	22 (29.7)	49 (66.2)
3. The company provides attractive career paths to internal auditors*	10 (13.5)	43 (58.1)	21 (28.4)
4. A formal quality assurance program exists for the internal audit function*	34 (47.2)	26 (36.1)	12 (16.7)
5. The internal audit function is structured for the efficient accomplishment of audit objectives	3 (4.1)	28 (38.4)	42 (57.5)
6. The internal audit function is staffed with qualified internal auditors	4 (5.5)	35 (48.0)	34 (46.5)
7. The internal audit function assigns responsibility to staff based on their qualifications.	1 (1.4)	23 (31.5)	49 (67.1)
8. The internal audit function uses Professional Standards of the IIA as audit evaluation criteria*	17 (23.3)	34 (46.5)	22 (30.1)

(continued)

Table 1. (Continued)

		Not at All	To some Extent	To a Great Extent
9.	The internal audit function encourages staff to include specific procedures to search for instances of fraudulent of financial reporting	9 (12.3)	52 (71.2)	12 (16.4)
10.	The internal audit function considers the experience of staff an important criterion in making assignments	1 (1.4)	25 (34.2)	47 (64.4)
11.	The internal audit function provides a strong sense of professionalism*	3 (4.1)	15 (20.5)	55 (75.3)
Objectivity and Independence				
12.	The audit committee reviews changes in the role of your internal audit function*	5 (6.8)	44 (60.3)	24 (32.9)
13.	The internal audit function considers objectivity an important criterion when assigning staff to audits	1 (1.4)	9 (12.3)	63 (86.3)
14.	The internal audit function reports findings directly to parties who can take corrective action*	0 (0.0)	4 (5.5)	69 (94.5)
15.	The internal audit function has formal procedures to determine if an internal auditor lacks independence with respect to an auditee	27 (37.0)	36 (49.3)	10 (13.7)

(continued)

Table 1. (Continued)

	Not at All	To some Extent	To a Great Extent
Nonfinancial Audit Findings			
16. The internal audit function considers the effect of non-financial audit findings on the fairness of the company's financial statements*	2 (2.7)	25 (34.2)	46 (63.0)
Involvement in the External Audit			
17. The internal audit function coordinates its role in the audit of the consolidated financial statements*	20 (27.4)	42 (57.5)	11 (15.1)

Notes: Numbers in parentheses represent total percentage of respondents.
*Indicates a recommendation of the Treadway Commission (NCFFR 1987).

Effectiveness of the Internal Audit Function

Based on the responses shown in Table 1, it appears that most of the measures related to the effectiveness of the internal audit function have been implemented, at least to some extent, by a large percentage of the sample companies. This is not surprising, because public companies directly benefit to the extent that more effective internal audits are conducted. Of the 11 effectiveness measures shown in Table 1, 7 (numbers 1, 2, 5, 6, 7, 10, and 11) were cited by approximately one-half of the responding companies as being implemented "to a great extent." In addition, the least implemented of these 7 measures, staffing the internal audit function with qualified internal auditors, still had been implemented to at least some extent by 94.5% of the companies.

Of the four remaining measures relating to the effectiveness of the internal audit function, two appear to have not been implemented by a relatively large number of sample companies. These measures are the existence of a formal quality assurance program (number 4) and the use of the IIA Standards as evaluation criteria (number 8).

These measures are somewhat related, as the IIA's Standards indicate that "[t]he purpose of this [quality assurance] program is to provide reasonable assurance that audit work conforms with these Standards" (IIA 1978, Section 560.01). While the IIA's Standards require internal audit directors to establish and maintain a quality assurance program, almost one-half (47.2%) of the respondents indicated that their companies had not done so. Possible explanations for the absence of a quality assurance program for such a large number of companies are the costs associated in implementing the program as well as proprietary issues. Qualified individuals who could perform a quality review would come primarily from two sources: CPA firms and the internal auditing departments of other companies. The work performed by the former individuals would be costly and the latter individuals would be privy to company secrets; neither situation appears to be appealing.

Interestingly, over 23% of the sample respondents indicated that their internal audit functions did not consider the IIA Standards as performance evaluation criteria. This may result from one of two factors. First, the IIA Standards provide only general guidance to internal auditors regarding the scope and performance of audit work. For example, when addressing evidence-gathering procedures, the IIA Standards indicate that information should be sufficient, competent, relevant, and useful (IIA 1978, Section 420.02); however, criteria are not provided as to what constitutes sufficient, competent, relevant, and useful evidence. Second, unlike external auditors, internal auditors are not subject to litigation from external parties for performing substandard work. As a result, the need to adhere to a set of published guidelines or standards is lessened. The failure of such a high percentage of companies to consider the IIA's Standards in this manner is an interesting question that deserves further attention.

Two additional measures appear to have been only partially implemented by a large number of companies: providing attractive career paths to internal auditors (number 3) and encouraging staff to include specific procedures to search for instances of fraudulent financial reporting (number 9). Based on the high number of respondents indicating that their companies only encourage staff to include procedures to search for fraud to some extent (71.2%), it appears that companies should expand the roles of their internal auditors to focus on fraudulent financial reporting. Also, the IIA may consider revising its Standards to explicitly mention fraudulent financial reporting as a responsibility of the internal audit function. The responses for providing attractive career paths to internal auditors indicate some current level of dissatisfaction with internal auditing as a long-term career. As a result, companies should consider taking measures such as providing increased financial rewards and increased upward mobility to internal auditors within the organization.[5]

Objectivity and Independence

Of the four measures related to objectivity and independence, companies appear to have implemented two of these to a lesser extent. While a large percentage (93.2%) of the sample companies' audit committees review changes in the role of the internal audit function, this activity is only partially implemented by most of these companies (60.3%). Because the internal audit function is an important resource for the audit committee, it appears that companies should increase the extent of involvement by their audit committees. The importance of audit committee involvement on the objectivity and independence of the internal audit function is illustrated by recommendations from both the Treadway Commission and the IIA's Standards (1978, Section 110).

The most troubling finding relating to objectivity and independence is the relatively large number of companies having no formal procedures to identify instances where independence may be lacking (27 companies, or 37%). The importance of independence is explicitly acknowledged by its incorporation in the IIA's Standards. While these Standards require companies to consider independence in assigning their internal auditors, they provide no guidance on formally documenting independence or considering independence-related issues. By adopting formal documentation procedures, such as those required by public accounting firms under the AICPA's quality review program, companies should be more likely to consider and identify possible conflicts of interest that exist between internal audit staff and the auditee.

Nonfinancial Audit Findings and Involvement in the External Audit

The final two categories of suggestions relate to considering the effect of nonfinancial audit findings on the fairness of the company's financial statements and coordinating the work of the internal audit function with that of the external auditors. Respondents indicated that the former of these measures has been implemented by almost all of the sample companies; only 2.7% of the internal audit managers indicated that nonfinancial audit findings are not considered in evaluating the fairness of the company's financial statements. In addition, most of the respondents indicated that nonfinancial audit findings are considered to a great extent (63.0%).

In contrast, it appears that opportunities for improvement exist with respect to coordinating the role of the internal auditor in the independent audit engagement. Only 11 companies (15.1%) indicated that their companies coordinate the efforts of their internal audit functions with the external auditors' examinations to a great extent; almost twice as many (27.4%) responded that their companies have not done so. The Treadway Commission's report indicates that, because of their superior knowledge of the organization and its internal control structure and their involvement at the division level, coordinating the efforts of internal auditors with external auditors has the potential to reduce the incidence of fraudulent financial reporting. This potential is further indicated by recent pronouncements by both the IIA (*Statement on Internal Auditing Standards No. 5*) (IIA 1987) and the AICPA (*Statement on Auditing Standards No. 65*) (AICPA 1991) on the relationship between internal and external auditors.

Clearly, companies should consider increasing the extent of involvement of their internal audit functions in the external audit and coordinating the efforts of the internal auditor with those of the external auditor. One possible explanation for their failure to do so was the uncertainty surrounding the AICPA's position on this issue prior to the issuance of *Statement on Auditing Standards No. 65* (AICPA 1991). Prior to this pronouncement, external auditors were provided with only limited guidance on considering the work of the internal audit function. The recent issuance of this pronouncement may result in external auditors requesting increased participation by the internal audit function.

Effect of Manager and Company Characteristics on Implementation

To determine whether the implementation of the measures shown in Table 1 was influenced by the (1) certification of the internal audit manager, (2) experience of the internal audit manager, or (3) size of the sample company, chi-square tests were conducted. Internal audit managers were classified as being either certified or noncertified based on the possession of the CIA, CPA, or CMA designation. Of the 74 respondents, 45 were CPAs and 13 were CIAs. For experience and size, observations were split into two categories based on median levels of experience and size. Because the small number of responses falling in the "not at all" category in Table 1 resulted in many cells having fewer than 5 expected observations, this category was combined with the "to some extent" category for purposes of analysis.[6]

The results of the chi-square analysis indicated that the certification of the internal audit manager influenced the implementation of measures to reduce the incidence of fraudulent financial reporting more than his or her level of experience. Of the 17 measures in Table 1, 4 (providing attractive career paths, staffing the internal audit function with qualified auditors, using the IIA Standards as evaluation criteria, and providing a strong sense of professionalism) were implemented to a greater extent when the internal audit manager was not certified than when he or she was certified ($\alpha =$ 0.05). In contrast, only one significant difference existed for different levels of experience (the structuring of the internal audit function for the efficient accomplishment of objectives).

Interestingly, only the implementation of measures related to the effectiveness of the internal audit function significantly differed for companies with internal audit managers possessing different characteristics. In contrast, a much larger number of differences were noted between the implementation of the measures shown in Table 1 for smaller and larger companies. Not surprisingly, larger companies were more likely to implement the measures summarized in Table 1 than smaller companies. Two explanations may account for this finding. First, implementing many of the measures shown in Table 1 appears to be costly and larger companies may be better able to bear the cost of implementing many of these measures than smaller companies. Also, the larger sizes of many of these companies may make certain measures (i.e., setting forth the responsibility of the internal audit department in writing, structuring the internal audit function for the efficient accomplishment of objectives, and determining whether an internal auditor lacks independence with respect to an auditee) more important. Based on the results of the chi-square tests, a total of 8 effectiveness measures (1, 2, 3, 4, 5, 6, 8, and 11) and 3 objectivity measures (12, 13, and 15) were implemented to a greater extent by larger companies than smaller companies (at $\alpha = 0.05$).

Effects of Potential Legislation on Fraudulent Financial Reporting

In addition to identifying the extent to which various measures have been implemented by sample companies to reduce the incidence of fraudulent financial reporting, this survey also requested internal audit managers to evaluate the potential impact of ten forms of legislation on fraudulent financial reporting. These items were identified by reviewing professional literature, including those items in the pretest, and soliciting additional items and ideas from the profession. The responses of the internal audit managers are summarized in Table 2.

Based on the responses in Table 2, it appears that several of the forms of legislation examined in this survey are perceived to be effective in reducing the incidence of fraudulent financial reporting. Four of the ten items were evaluated as being either "moderately" or "very" effective in reducing the incidence of fraudulent financial reporting by at least 40% of the sample respondents. Internal audit managers felt that requiring internal audit directors to report directly to the audit committee would have the greatest impact on detecting and preventing fraudulent financial reporting. Of the 73 responses, over 50% (37) indicated that this requirement would be at least moderately effective in this regard. The perceived effectiveness of the legislation is consistent with the Treadway Commission's recommendation that the chief internal auditor report to a senior officer not involved in preparing the company's financial statements.

Two other forms of legislation that would expand the reporting responsibilities of internal audit directors were also perceived as relatively effective in preventing and detecting fraudulent financial reporting. These include reporting on (1) the strength of the organization's internal controls and (2) the existence and resolution of fraudulent financial transactions. These forms of legislation were viewed as at least moderately effective by 41.1% and 41.9% of the internal managers, respectively. These forms of legislation are similar in nature to a recommendation made by the

Table 2. Extent to Which Fraud Detection and Prevention Would be Enhanced by Proposed Legislation

	Not at All	Slightly	Moderately	Very
The director of internal audit should be given officer status and incur personal liability in any litigation involving fraudulent financial reporting	28 (38.9)	20 (27.8)	14 (19.4)	10 (13.9)
Internal auditors should be held liable for negligent nondiscovery of fraudulent transactions	27 (37.0)	23 (31.5)	13 (17.8)	10 (13.7)
Internal auditors should bring any suspicions of fraudulent financial reporting to the attention of the external auditor	23 (31.1)	29 (39.2)	15 (20.3)	7 (9.5)
The director of internal audit should report directly to the audit committee	14 (19.2)	22 (30.1)	19 (26.0)	18 (24.7)
The director of internal audit should express an opinion on the strength of internal controls	16 (21.9)	27 (37.0)	23 (31.5)	7 (9.6)
The director of internal audit should report on the existence of fraudulent financial transactions	14 (18.9)	29 (39.2)	25 (33.8)	6 (8.1)
Any internal auditor who "blows the whistle" about fraud should be protected from the media	19 (25.7)	26 (35.1)	17 (23.0)	12 (16.2)

(continued)

Table 2.   (Continued)

	Not at All	Slightly	Moderately	Very
Any internal auditor who "blows the whistle" should be rewarded financially	42 (56.8)	17 (23.0)	11 (14.9)	4 (5.4)
Certified internal auditors should be licensed to issue an opinion on internal controls	31 (42.5)	22 (30.1)	15 (20.5)	5 (6.8)
An Internal Control Standards Board should be created to issue standards on internal controls	23 (31.1)	20 (27.0)	17 (23.0)	14 (18.9)

Note: Numbers in parentheses represent total percentage of respondents.

Treadway Commission for increased reporting of management and audit committee activities to the public (NCFFR 1987). Consistent with these recommendations, recent legislation has been proposed to require companies to report on their internal control structures ("AICPA backs bill …" 1990) and immediately disclose any identified illegal acts to the SEC ("Bill would require …" 1991); at the current time, this legislation has not been approved by Congress. In addition, increased management reporting on the strength of the organization's internal control structure is consistent with recent guidance provided to external accountants by *Statement on Standards for Attestation Engagements No. 2* (AICPA 1993) for reporting on assertions made by management in their reports on internal control structure.

Finally, 41.9% percent of respondents felt that establishing an oversight board to issue standards on internal control structure (similar in nature to the role of the FASB for financial reporting) would be either moderately or very effective in reducing the incidence of fraudulent financial reporting. Interestingly, subsequent to the Treadway Commission's report, the Committee of Sponsoring Organizations (COSO) has issued a report that integrates various internal control concepts (COSO 1992). The purpose of this report is to provide organizations with an understanding of how they can design and maintain effective internal controls. However, it appears that internal audit managers felt that this report was merely a first step and that a full-time standards board would be beneficial to reduce the incidence of fraudulent financial reporting.

The remaining items shown in Table 2 were viewed as less effective by internal audit managers in preventing or detecting fraudulent financial reporting. Three of these items (the director of internal audit should incur personal liability in litigation involving fraudulent financial reporting, internal auditors should be held liable for negligent nondiscovery of fraudulent transactions, and internal auditors should be financially rewarded for "blowing the whistle") would apparently provide greater incentives to internal auditors for detecting fraudulent financial reporting. It appears that internal audit managers do not feel that incentives (either financial rewards or monetary fines and / or reputation losses through litigation) are needed to encourage internal auditors to detect instances of fraudulent financial reporting. This may be due to a belief that professionals should not require additional incentives in order for them to accomplish their assigned responsibilities. This, coupled with incremental exposure to liability and government regulation, may explain the low ratings in these categories.

In addition to the above, it does not appear that (1) communicating suspicions of fraudulent financial reporting to the external auditor, (2) protecting internal auditors who "blow the whistle" about fraud from media exposure (and, possibly, losing their jobs), and (3) licensing certified internal auditors to report on internal controls are perceived as effective in reducing fraudulent financial reporting. The results for communicating suspicions to the external auditor are not consistent with the Treadway Commission's recommendation that the internal auditors' involvement is properly coordinated with the independent public accountant's examination. Apparently, internal audit managers do not feel that bringing possible instances of fraudulent financial reporting to the independent auditor's attention is particularly beneficial.

Similar to the responses in Table 1, chi-square tests were performed to determine whether internal audit managers' perceptions of the effectiveness of the potential legislation shown in Table 2 were influenced by their experience, certification, or company size. For purposes of this analysis, the responses "not at all" and "slightly" were combined into one category; the responses "moderately" and "very" were combined into a second category. As before, this combination was necessitated by small expected cell frequencies for some categories.

As with the measures to reduce the incidence of fraudulent financial reporting (see Table 1), only a small number of significant differences were noted based on characteristics of the internal audit manager. Two of the proposed items of legislation shown in Table 2 were perceived to be more effective by internal auditors who were certified: licensing CIAs to issue an opinion on internal controls and creating an internal control standards board ($\alpha = 0.05$). In contrast, no significant differences were noted in the perceived effectiveness of the ten potential forms of legislation for more and less experienced internal audit managers.

A larger number of differences were noted between the perceived effectiveness of the proposed legislation shown in Table 2 for internal audit managers affiliated with larger and smaller companies. Interestingly, internal audit managers affiliated with smaller companies felt that proposed legislation would be more effective in reducing the incidence of fraudulent financial reporting than the internal audit managers affiliated with larger companies.[7] While it is not clear why this finding resulted, it is possible that larger companies may have already taken steps to reduce the incidence of fraudulent financial reporting. If so, many of the forms of legislation shown in Table 2 may not be considered as effective by the audit managers of these companies. In other words, if smaller companies have not implemented measures to reduce the incidence of fraudulent financial reporting, proposed legislation may be perceived as more effective by internal audit managers affiliated with these companies. This explanation is consistent with the earlier conclusion that larger companies have implemented measures to reduce the incidence of fraudulent financial reporting to a greater extent than smaller companies.

CONCLUSIONS

This study surveyed internal audit managers to determine (1) the extent to which various measures have been implemented by public companies to reduce the incidence of fraudulent financial reporting and (2) the perceived effectiveness of various forms of legislation in reducing the incidence of fraudulent financial reporting. The results of this survey revealed that, in general, public companies have implemented the Treadway Commission's recommendations to at least some extent. Ancillary analysis revealed that measures to reduce the incidence of fraudulent financial reporting have been implemented to a greater extent by larger companies and companies whose internal audit managers do not possess the CIA designation.

Notable exceptions to the companies' implementation of measures to reduce the incidence of fraudulent financial reporting relate to the existence of formal quality assurance programs and the relatively large number of companies having no formal procedures to identify instances where independence may be lacking. The Professional Standards Committee of the IIA is responsible for establishing auditing standards for internal auditors. This committee may wish to investigate various explanations for the failure of companies to implement these programs in improving the performance of the internal audit function and its ability to detect fraudulent financial reporting.

Internal audit managers' perceptions of the effect of various forms of legislation on the incidence of fraudulent financial reporting revealed that four measures were perceived to be relatively effective in this regard. These include: (1) having the director of internal audit report directly to the audit committee, (2) requiring the director of internal audit to issue an opinion on internal control structure, (3) requiring the director of internal audit to report on the existence of fraudulent financial transactions, and (4) creating a board that issues standards related to internal control structure. In general, proposed forms of legislation were perceived to be more effective in reducing the incidence of fraudulent financial reporting by internal audit managers affiliated with smaller companies than with those affiliated with larger companies. While requiring specific communications by the director of internal audit comes under the purview of the Institute of Internal Auditors, an internal control standards board affects all members of COSO. Because internal control effects may be viewed differently by internal auditors, management, and external auditors, the feasibility of such a project must be investigated by the entire coalition.

The findings of this survey have implications for public companies, their internal audit functions, and oversight bodies (such as the Securities and Exchange Commission). Public companies should consider implementing the recommendations of the Treadway Commission and some of the other measures examined in this study to reduce the incidence of fraudulent financial reporting. Some of these measures (e.g., implementing a quality assurance program, using the IIA Standards as evaluation criteria, and specifically searching for instances of fraudulent financial reporting) will also impact the operations and responsibilities of these companies' internal audit functions. Finally, the perceived effectiveness of different types of legislation in reducing the incidence of fraudulent financial reporting suggests that the SEC consider increasing the reporting requirements for internal audit directors to include reports on the organization's internal control structure and the existence and resolution of fraudulent financial transactions. Based on the internal audit managers' responses, it appears that doing so will enable public companies to enhance the integrity of their financial reporting processes.

ACKNOWLEDGMENT

The authors would like to acknowledge the helpful comments of the editor and reviewer.

NOTES

1. See Savage (1988) for a summary of these recommendations.
2. Bull (1991) and Bull and Sharp (1989) discuss the recommendations of the Treadway Commission as they affect public companies' upper management and audit committee, respectively.
3. Currently, these standards address five major areas: (1) independence, (2) professional proficiency, (3) scope of work, (4) performance of audit work, and (5) management of the internal auditing department.
4. Our sample was limited to publicly held companies located in California because two of the authors' academic affiliations enhanced the probability of achieving an adequate response rate. As our sample was not limited to specific industries, we feel that the results should be generalizable to other publicly held companies in the United States. While we are unaware of any systematic differences between publicly held companies operating in California and those operating in other areas of the United States, any such differences may affect generalizability of our results.

5.   One limitation in interpreting the results for this question is that internal auditors may consider their own self-interest in responding. If internal auditors indicated that their employers provided attractive career paths "to a great extent," companies would not be as likely to increase the attractiveness of the internal auditing work environment. As a result, for this question, respondents may feel it more beneficial to report a less favorable response (e.g., "not at all" or "to some extent"). This limitation should be considered in evaluating the responses to this and other questions contained in the research instrument. For example, internal audit managers may be reluctant to admit that their internal audit function is not staffed with qualified internal auditors.

6.   Thus, two categories were compared in conducting chi-square analyses: "to a great extent" (the third column in Table 1) and a combination of the first two columns in Table 1. Cochran (1954) notes that no more than 20% of the expected cell frequencies in any chi-square analysis should contain fewer than 5 expected observations. Classifying responses in the above manner does not result in any violations of this guideline. It should be noted that this method of analysis still allows us to determine whether internal audit manager experience, internal manager certification, and company size influence the extent to which the measures summarized in Table 1 have been implemented by sample companies.

7.   The internal audit managers' perceptions of the effectiveness of the following forms of legislation were found to significantly differ (at $\alpha = 0.10$): (1) holding internal auditors liable for negligent nondiscovery of fraudulent transactions, (2) bringing any suspicions of fraudulent financial reporting to the attention of the external auditor, (3) expressing an opinion on the strength of internal controls, (4) protecting internal auditors who "blow the whistle" from the media, (5) financially rewarding internal auditors for "blowing the whistle," and (6) creating an internal control standards board.

REFERENCES

AICPA backs bill requiring reports on internal controls. 1990. *The CPA Letter* (October): 1.

American Institute of Certified Public Accountants. 1991. *Statement on Auditing Standards No. 65*, The Auditor's Consideration of the Internal Audit Function in an Audit of Financial Statements. New York: AICPA.

_____. 1993. *Statement on Standards for Attestation Engagements No. 2*, Reporting on an Entity's Internal Control Structure Over Financial Reporting. New York: AICPA.

Bill would require notification of illegal acts to SEC. 1991. *The CPA Letter* (September): 1.

Bull, I. 1991. Board of director acceptance of Treadway responsibilities. *Journal of Accountancy* (February): 67-74.

Bull, I., and F.C. Sharp. 1989. Advising clients on Treadway audit committee recommendations. *Journal of Accountancy* (February): 46-52.

Cochran, W.G. 1954. Some methods for strengthening the common X^2 Test. *Biometrics*: 417-451.

Committee of Sponsoring Organizations of the Treadway Commission (COSO). 1992. *Internal Control—Integrated Framework*. New York: COSO.
Institute of Internal Auditors. 1978. *Standards for the Professional Practice of Internal Auditing*. Altamonte Springs, FL: IIA.
_____. 1987. *Statement on Internal Auditing Standards No. 5*, Internal auditors' relationship with independent outside auditors. Altamonte Springs, FL: IIA.
National Commission on Fraudulent Financial Reporting (NCFFR). 1987. *Report of the National Commission on Fraudulent Financial Reporting*. Washington, DC: NCFFR.
Savage, L.J. 1988. Special report: National Commission on Fraudulent Financial Reporting. *Internal Auditor* (December): 54-61.

AN EMPIRICAL ANALYSIS OF THE COMPARABILITY OF DISCLOSURE TENDENCIES WITHIN AND ACROSS INDUSTRIES:

THE CASE OF HAZARDOUS WASTE LAWSUITS

Philip Little, Michael Muoghalu, and David Robison

ABSTRACT

This paper extends the work of Thompson, Smith, and Williams (1990) on the adequacy of loss contingency disclosures by examining the financial statement treatment of hazardous waste lawsuits. The extension is in three directions. First, we control for the relative importance of suits using an event-study method to estimate the impact of the announcement of the suit on stock prices. We assume large negative abnormal returns signal suits important to investors. Second, we separate firms in the sample into subgroups based on industry. Third, we examine a single type of suit—hazardous waste

Research in Accounting Regulation, Volume 9, pages 105-117.
ISBN: 1-55938-883-8

lawsuits—rather than suits in general. Our findings reveal substantial variation in the financial statement disclosures of hazardous waste lawsuits across industries even when controlling for the expected magnitude of the loss. Thus, like Thompson, Smith, and Williams, we find that financial statement users may have difficulty making comparisons across firms with outstanding loss contingencies. Further, because we focused on a single type of lawsuit and controlled for the importance of the suits, our results suggest more strongly that there is a lack of comparability in lawsuit disclosures especially across different industries.

INTRODUCTION

Thompson, Smith, and Williams (1990) report that the ambiguities of the wording in *Statement of Financial Accounting Standards* (*SFAS*) *No. 5*, coupled with managements' possible reluctance to disclose bad news, such as the details of outstanding litigation, lead to disclosures that are often "general, vague, or incomplete." Thompson, Smith, and Williams' findings are important to accountants and financial statement users for at least two reasons. First, there are the theoretical and practical questions of the comparability of financial statements. While in theory *SFAS 5* was to provide for some consistency in disclosures of loss contingencies, in practice, inconsistent treatment might arise for a variety of reasons discussed below. In both theory and practice, inconsistent treatment of loss contingencies could lead to the diminution of the value of financial statements over time. Second, firms choosing not to disclose suits in the financial statements may be risking failure-to-disclose lawsuits if the original suits subsequently lead to material losses.

This paper extends the work of Thompson, Smith, and Williams on the adequacy of loss contingency disclosures by examining whether the inconsistencies found by Thompson, Smith, and Williams can be explained by controlling for additional factors. In particular, we control for the importance of the suits, the industry a firm is in, and the type of suit. Because these factors are likely to affect disclosure decisions, inconsistencies found after controlling for these factors will provide substantial support for the Thompson, Smith, and Williams results.[1]

We control for the relative importance of suits using an event-study method to estimate the impact of the announcement of the suit on stock prices. Without a control for the importance of suits, it is possible that any apparent inconsistency in the financial statement treatment of lawsuits could be the result of the different sizes of the suits. If all large important suits are fully disclosed and discussed while unimportant ones are dismissed as immaterial, the financial statement user is getting appropriate information in spite of the appearance of inconsistent amounts of information being released.

Inconsistent treatment may also be an industry-related phenomenon. It is possible that some industries provide full disclosure while others systematically fail to disclose suits. Thus, the apparent inconsistency may occur only when making comparisons across industries. Cowen, Ferreri, and Parker (1987) and Dierkes and Preston (1977) suggest reasons why particular industries might have higher disclosure rates than others. Because of the potential importance of industry differences, we separate our sample of firms into two specific industry subgroups—waste management and petrochemicals—and a catch-all "Other Industries" group.[2]

Inconsistent treatment of suits may also result from the inherent differences in the types of suits being examined. Different types of suits (antitrust, product liability, wrongful dismissal, hazardous waste, etc.) would have different potential damages, burdens of proof, probabilities of losing, and potentials for attracting additional suits. By selecting only hazardous waste lawsuits, we expect greater consistency in how the suits are treated. Hazardous waste lawsuits were selected as the type of loss contingency to be examined in the current study for two reasons. First, hazardous waste lawsuits have a high potential for material impacts in comparison to other types of suits. Given that hazardous waste suits have sought clean-up costs and damages up to $2 billion and even relatively small cases can involve several million dollars, the potential for material impacts is quite high (*WSJ*, 1983). Second, hazardous waste suits have a large variation in both the size of the damages being sought and the size of firms being sued. Given this variation, we expect substantial variation in the importance of the suit, making it necessary (and possible) to control for this variable.

DISCLOSURE REQUIREMENTS

Financial statement disclosure requirements for loss contingencies such as hazardous waste lawsuits are provided by *SFAS 5*. A loss contingency must be recorded as a loss on the income statement and as a liability on the balance sheet if it is probable (likely to occur) that an asset has been impaired, or a liability incurred, and the amount of the loss is material and can be reasonably estimated. No disclosure is required if the likelihood of a loss is considered remote or the lawsuit is immaterial. If the likelihood of a material loss is reasonably possible (between remote and probable), a footnote disclosure is required indicating the nature of the loss contingency. Ambiguities arise because the terms "probable," "remote," and "reasonably possible" are defined using words like "likely" and "slight" rather than numerical probabilities. As pointed out by Thompson, Smith, and Williams, the lack of specific probability ranges allows for considerable professional discretion in disclosure decisions. This ambiguity, combined with the management's reluctance to admit the bad news of a probability or possibility of loss, may result in a lack of comparability in disclosures of hazardous waste lawsuits.

ENVIRONMENTAL LAWS

All suits in our sample were brought under either the Resource Conservation and Recovery Act (RCRA) or the Comprehensive Environmental Response Compensation and Liability Act (Superfund). The objective of the RCRA is to limit environmental damage by providing a system for controlling hazardous wastes from their creation through final disposal, commonly referred to as "cradle to grave" control. The RCRA, passed in 1976 and amended in 1984, assigns the Environmental Protection Agency (EPA) the job of creating waste handling and disposal standards, and establishing a manifest system to track the flow of wastes. Although the emphasis of the RCRA is defining legal disposal practices, it allows for EPA intervention and lawsuits when a site presents "an imminent and substantial danger to health or the environment."

The Superfund Act, passed in 1980, is currently the primary mechanism for governmental response to releases of hazardous materials into the environment and for cleanup of abandoned sites. While the Superfund Act may have a deterrent effect (Muoghalu, Robison, and Glascock 1990), its focus is the clean up of materials previously released in the environment rather than the prevention of additional releases. When the government responds to a Superfund site, it is required to attempt to recover all costs associated with the response and remedial actions as well as claims for damages to natural resources.

In addition to governmental suits, both the RCRA and Superfund Acts have provisions that permit individuals or groups to bring suits against firms. However, the RCRA limits individuals to seeking only injunctive relief and attorneys' fees. The Superfund Act expanded the right to sue to permit plaintiffs to seek compensation for damages caused by hazardous substances, a change that encourages more individuals to bring more suits. For example, suppose the Environmental Protection Agency sues a firm because hazardous wastes have contaminated local ground water. Individuals whose wells were contaminated could bring suits seeking compensation for the loss of the wells, the costs of obtaining clean water, the loss in property value, exposure to the chemicals, and psychological and emotional damages. Given the potential for additional damage suits, firms would have an incentive not to disclose the suit in the financial statements.

Both the RCRA and Superfund Acts establish strict, joint, and several liability. Under the joint and several liability rulings, firms can be held financially responsible well beyond their contribution to a site. Sole responsibility can occur if no other contributors to the site can be found, or if other contributors cannot pay their share of the costs. Obviously, the joint and several rulings make the evaluation of a suit resulting from a multiple-user site more complex.

DISCLOSURE INCENTIVES

There are several factors which may affect disclosure decisions. First, there is the understandable reluctance on the part of a company's management to disclose bad news. When the costs of disclosure are low (likelihood of additional suits is low), firms are more likely to disclose the suit immediately than when the costs of disclosure are high. Thus, an initial failure to disclose might represent a delay in confirming bad news about the firm rather than an indication that the suit is not material. If firms are using nondisclosure as a delaying tactic, we would expect inconsistent disclosure even after controlling for the importance of the suit. Both Little, Muoghalu, and Robison (1995) and Thompson, Smith, and Williams discuss the possibility that disclosure of bad news, such as a lawsuit, might hurt the firm's chances of winning the suit. Disclosing a suit is an admission that a loss is at least reasonably possible, an admission which firms will be reluctant to make if it affects the probability of successfully defending against the suit.[3]

Second, as discussed in Rigsby, Lambert, and Alexander (1989), high complexity cases are perceived by managers and audit partners as more material than low complexity cases, making disclosure more likely. For hazardous waste suits, complexity is caused by difficulty in assessing the type and concentration of the chemicals, the amount of chemicals dumped, the extent of the spread of the chemicals, clean-up costs, and magnitude of the liability in the case of multiple-user sites. Given the complexity of most hazardous waste disposal cases, the Rigsby et al., results would suggest a general bias towards disclosing hazardous waste lawsuits.

A third factor affecting disclosure is uncertainty. Thompson, Smith, and Williams suggest that management may be less inclined to disclose information concerning matters that involve a greater degree of uncertainty. Presumably firms with enhanced ability to estimate the probability of loss and assess a reasonable estimate of the loss would be more likely to disclose a suit. Hazardous waste disposal firms, because of their control of sites and the record keeping requirements of the RCRA, would seem to have less uncertainty than other firms. Similarly, because petrochemical firms are more likely to use on-site disposal (Magorian and Morell 1982), the degree of uncertainty should be lower than for firms in our Other Industries category.

Fourth, as suggested in Cowen, Ferreri, and Parker (1987), industries prone to governmental regulatory pressures are more likely to disclose socially sensitive matters. Similarly, Dierkes and Preston (1977) suggest that firms which have "high visibility" are more likely to disclose, because they are aware of the public or private monitoring of their actions. Thus, the visibility of firms in the petrochemical industry should be associated with a higher disclosure rate, irrespective of the size of the abnormal returns.

Finally, firm size may affect disclosure decisions. According to Trotman and Bradley (1981), firm size should be associated with an increased likelihood of disclosure because large corporations receive more public attention and pressure. However, other things being equal, an increase in the size of firms will decrease the materiality of a given suit which should reduce the probability of disclosure.

DATA

The hazardous waste lawsuits examined in this study were selected from the *Wall Street Journal Index*. In particular, lawsuits brought between 1977 and 1986 against companies for hazardous waste mismanagement were chosen. To be included in the sample, the following five requirements had to be met:

1. The lawsuit must have been filed under the RCRA or Superfund Acts and reported as involving hazardous waste mismanagement.
2. The lawsuit must have been the first brought against a firm at a particular location. Subsequent suits associated with a site are excluded.
3. The firm being sued had to be listed on the New York or American Stock Exchange so that stock returns would be available from the Center for Research in Security Prices (CRSP) database.
4. Firms had to have no other significant announcement in the 10 days prior to and 10 days following the announcement of the lawsuit.
5. Financial statements for the firms being sued had to be available through the Disclosure Inc. Service.

A total of 103 lawsuits against 58 firms met all five conditions and are included in the sample. In addition, the financial statements for each company facing the hazardous waste lawsuits were examined to see if a footnote disclosure (required for firms if loss is probable or reasonably possible) was provided in the year that the suit was filed. If the nature of the individual lawsuits were specifically mentioned in the footnotes, they were placed in the yes disclosure category and if not, they were placed in the no disclosure category.

Table 1.   Number of Sample Lawsuits Filed by Year

1977	1978	1979	1980	1981	1982	1983	1984	1985	1986
2	2	4	19	5	22	24	7	9	9

Table 1 presents the number of lawsuits filed by year, while Table 2 presents the distribution of the suits across the three different industry types. As can be seen in Table 2, 55 of the lawsuits were filed against petrochemical firms, 10 of the lawsuits were filed against waste management firms, and 38 of the lawsuits were filed against companies in a wide variety of other industries that are not normally associated with hazardous waste handling.

EVENT-STUDY METHOD

Comparison of the importance of hazardous waste lawsuits across firms and industries is difficult because of the specific characteristics of each suit. The dollar amount sought fails to provide a consistent estimate of the importance of suits because the actual payments by firms are frequently unrelated to the amount sought. Settlement values cannot be used because most out-of-court settlements contain nondisclosure agreements.

Based on standard financial market theory, an unbiased estimate of the importance of a suit can be found in the market reaction to the announcement of a lawsuit. We therefore use a standard event-study technique to estimate the impact of a suit on the present value of the future cash flows to stockholders. In using the abnormal returns from an event-study as a measure of the importance of a suit, we are assuming that the market can evaluate the probability of the firm losing the suit, the size of the damage awards, the loss of goodwill, and the firm's insurance coverage.

Table 2.   Distribution of Sample Lawsuits by Industry Type

Industry Type	Number of Lawsuits	Number of Firms Sued
Petrochemical	55	24
Waste Management	10	4
Other Industries	38	30
Total	103	58

Following Little, Muoghalu, and Robison, we use a modified version of the Dodd and Warner (1983) event-study technique. We treat the date that each lawsuit is first reported in the print media as the event date, day 0, and all other days are measured relative to day 0. The market model is estimated for each firm over a 200-day interval, day -261 to day -61. Abnormal returns for each firm were computed for a 121-day event window, days -60 to 60. The abnormal returns were then averaged across firms and cumulated over various time intervals with appropriate test statistics. Given the results of Little, Muoghalu, and Robison which found significant abnormal returns on only the interval -1 to 0, we use those abnormal returns as our measure of the importance of the individual suits.

Despite the careful date screening and proper statistical techniques, the event-study approach to measuring the importance of the hazardous waste lawsuits has two important limitations. First, because stock prices move randomly at times, not all movements that are measured in this study are necessarily caused by the suit announcements. Second, the information about suits made available at the time of the announcements is limited, leading investors to make estimates which may prove to be wrong as more information is released. If investors are using inaccurate information, the estimates of the importance of individual suits may be off in either direction. Thus, while the estimates presented below are unbiased, they remain subject to error which could affect the interpretations.

RESEARCH FINDINGS

As can be seen in Table 3, the petrochemical industry has the highest propensity to disclose suits, followed by the waste management industry and other industries. The petrochemical industry disclosed 34 out of 55 (almost two-thirds) of the suits versus a disclosure of only 14 out of 38 (about one third) of the suits in the other industry category and 5 out of 10 (one-half) in the waste management industry. A chi-square test of the data reveals that these differences are statistically significant at the .05 level. This pattern also holds true for abnormal losses in excess of 1%. For abnormal losses exceeding 3%, the other industry category disclosed one-half of the suits while the pattern for the petrochemical industry and the waste management industry was the same as in the other categories.

Table 3. Hazardous Waste Lawsuit Disclosure Tendencies
According to Industry Type and Size of Abnormal Returns

Industry Type	Footnote Disclosure	Number of Suits	Abnormal Losses* $< 1\%$	Abnormal Losses $\geq 1\%$	Abnormal Losses $\geq 3\%$
Petrochemical	Yes	34	21	13	6
	No	21	14	7	2
Waste	Yes	5	2	3	3
Management	No	5	2	3	3
Other Industries	Yes	14	10	4	3
	No	24	15	9	3

Note: *These abnormal losses were not only less than 1%, some were actually small but positive numbers.

The disclosure rates for each industry are fairly consistent across the range of abnormal returns. For example, firms in the petrochemical industry disclosed only 21 out of 35 suits even though the abnormal losses exceeded 1%. Thus, firms appear not to be making disclosure decisions on the basis of the market's perceived importance of the suit. In addition, none of the three industry groups appear to be systematically disclosing all suits or all important suits. Thus, while there are variations in the rate of disclosure across industries, we find no evidence that financial statement users can expect systematic treatment of hazardous waste lawsuits. Within the limitations of our technique, these results support and enhance those of Thompson, Smith, and Williams, and raise questions about the comparability of financial statements when firms have outstanding loss contingencies.

Table 4 presents additional data including the effects of firm size (total assets) and the size of the abnormal stock market returns (MCPE) on disclosure tendencies. Comparisons in Table 4 are made for both within- and between-industry comparisons. That is, is there a significant difference in the size of the firms or the size of the abnormal stock market returns between industries and between the yes and no disclosure categories within each industry type? Z-tests of the difference between means of total assets and abnormal stock market returns for the petrochemical and other industry category (both between and within) reveal no significant differences (highly insignificant). However, the difference between the means for total assets within the waste management industry were found to be

Table 4. Descriptive Statistics for Disclosed and Non-disclosed suits
(data from 10-K Reports)

Variable	Disclosed or Not	N^a	Mean	Standard Deviation
Petrochemical:				
Total Assets[b]	Yes	34	$10,204	$ 8,348
	No	21	$15,229	$19,712
MCPE[c]	Yes	34	-.006	.026
	No	21	-.006	.019
Waste Management:				
Total Assets[b]	Yes	5	$ 713	$ 784
	No	5	$1,707	$ 505
MCPE[c]	Yes	5	-.030	.049
	No	5	-.091	.142
Other Industries:				
Total Assets	Yes	14	$6,797	$12,026
	No	24	$6,526	$ 9,519
MCPE	Yes	14	-.013	.047
	No	24	-.003	.034

Notes: [a]Number of lawsuits
[b]In millions of dollars
[c]Estimates for each firm taken from the market model. Mean Daily Cumulative Prediction Errors (MCPE), in percent, for period of -1 to 0 days around the announcement of the hazardous waste lawsuit.

significant at the .01 level and the difference between the means for total assets between the petrochemical and waste management industries was found to be significant at the .05 level. Given the fact that the waste management industry is represented by only four companies, the significant differences may be sample specific.

CONCLUSIONS

Our findings reveal substantial variation in the financial statement disclosures of hazardous waste lawsuits across industries even when controlling for the type of industry and the expected magnitude of the loss. Thus, like Thompson, Smith, and Williams (1990), we find that financial statement users may have difficulty making comparisons across firms with outstanding loss contingencies. Further, because we focused on a single type of lawsuit and controlled for the importance of the suits, our results suggest more strongly that there is a lack of comparability in lawsuit disclosures especially across different industry types.

These results are, however, limited by the ability of the event-study technique to accurately estimate the importance of the suits.

The reported variation in the propensity to disclose details of outstanding lawsuits across industries is consistent with the hypothesis put forth by Cowen, Ferreri, and Parker (1987) and Dierkes and Preston (1977) that industries, such as the petrochemical industry, which are subject to greater governmental regulatory pressures regarding socially sensitive matters will be more likely to disclose information about those matters. Further, given the complexity introduced by the RCRA and Superfund laws and the ambiguity of the reporting requirements in *SFAS 5*, the variation is perhaps not-too-surprising. However, finding that the variation is consistent with these expectations does not change the fact that inconsistent disclosure patterns limit the ability of financial statement users to make comparisons across firms and industries.

NOTES

1. We gratefully acknowledge the helpful comments of two anonymous referees.

2. Other than the petrochemical and waste management industries, no industry had enough firms facing suits to create a significant industry grouping. The large majority of suits in the other category are against firms in manufacturing industries.

3. Some additional insights into possible motives for disclosing or not disclosing hazardous waste lawsuits can be found in the extensive discretionary disclosure literature. However, it is beyond the scope of this paper to fully explore each of these motives in detail. See Verrecchia (1983, 1990), Dye (1985a, 1985b, 1986), Gibbons, Richardson, and Waterhouse (1990), and Diamond (1985) for examples of this literature.

REFERENCES

Cowen, S., L. Ferreri, and L. Parker. 1987. The impact of corporate characteristics on social responsibility disclosure: A topology and frequency-based analysis. *Accounting, Organizations and Society* 12: 111-122.

Diamond, D. 1985. Optimal release of information by firms. *Journal of Finance* 40: 1071-1094.

Dierkes, M., and L. Preston. 1977. Corporate social accounting reporting for the physical environment: A critical review and implementation proposal. *Accounting, Organizations and Society* 2: 3-22.

Dodd, P., and J. Warner. 1983. On corporate governance: A study of proxy contests. *Journal of Financial Economics* 11: 401-438.

Dye, R. 1985a. Disclosure of nonproprietary information. *Journal of Accounting Research* 23: 123-145.

————. 1985b. Strategic accounting choice and the effects of alternate financial reporting requirements. *Journal of Accounting Research* 23: 544-574.

————. 1986. Proprietary and nonproprietary disclosures. *Journal of Business* 59: 331-366.

Financial Accounting Standards Board. 1975. *Statement of Financial Accounting Standards No. 5*, Accounting for Contingencies. Stamford, CT: FASB.

Gibbins, M., A. Richardson, and J. Waterhouse. 1990. The management of corporate financial disclosure: Opportunism, ritualism, politics, and processes. *Journal of Accounting Research* 28: 121-143.

Little, P., M. Muoghalu, and D. Robison. 1995. Hazardous waste lawsuits, financial disclosure, and investors' interests. *Journal of Accounting, Auditing, and Finance* 10: 383-398.

Magorian, C., and D. Morell. 1982. *Sitting Hazardous Waste Facilities*. Cambridge, MA: Ballinger.

Muoghalu, M., D. Robison, and J. Glascock. 1990. Hazardous waste lawsuits, stockholder returns and deterrence. *Southern Economic Journal* 57: 357-370.

Rigsby, J., K. Lambert, and E. Alexander. 1989. Experience and the quality of managerial decision making: The case of auditors' materiality judgments. *Journal of Managerial Issues* 1: 44-65.

Thompson, J., M. Smith, and J. Williams. 1990. An evaluation of the reporting standards for litigation: Some empirical evidence. In *Research in Accounting Regulation*, Vol. 4, eds. G.J. Previts, L. Parker, and O. Johnson, 43-57. Greenwich, CT: JAI Press.

Trotman, K., and G. Bradley. 1981. Associations between social responsibility disclosure and characteristics of companies. *Accounting, Organizations and Society* 6: 355-362.

Verrecchia, R. 1983. Discretionary disclosure. *Journal of Accounting and Economics* 5: 179-194.

————. 1990. Information quality and discretionary disclosure. *Journal of Accounting and Economics* 12: 365-380.

The Wall Street Journal. 1983. December 12, 4.

PART II

PERSPECTIVES

PUBLIC ACCOUNTING IN
AN OLDER SOCIETY:
SOME KEY PERSONNEL ISSUES

Stephen E. Loeb

ABSTRACT

This paper further develops existing literature relating the prospect
of an older population in America to personnel issues in public
accounting. Relevant legal considerations and ethical standards are
discussed. Policies relating to mandatory retirement of partners or
their equivalent in certain public accounting firms are considered. An
in-depth discussion of certain issues relating to the provision of
opportunities for older individuals is presented. The issue of "equity"
(Moody 1992, 208) between generations of individuals who work in
public accounting is considered.

Research in Accounting Regulation, Volume 9, pages 121-150.
Copyright © 1995 by JAI Press Inc.
All rights of reproduction in any form reserved.
ISBN: 1-55938-883-8

INTRODUCTION

This paper extends the existing literature on the relationship of the aging of America's population to personnel issues in the public accounting profession. Legal matters and ethical standards relating to age and personnel policies in public accounting firms are discussed. Existing mandatory retirement policies for partners or their equivalent of certain public accounting firms are noted. Selected issues relating to providing opportunities for elderly individuals in public accounting are considered in-depth. Also, the treatment of older members of the public accounting profession is considered in terms of the issue of "equity" (Moody 1992, 208) between generations of individuals in public accounting.

Population Projections

A number of authorities forecast the prospect of a longer living population in America comprised of relatively fewer younger individuals and more elderly individuals.[1] One recent set of population projections can be found in Spencer (1989, see particularly p. 1). He (1989, 1) indicates that "during the next 20 years, the elderly population ([which he defines as age] 65 and older) is projected to grow more slowly...." Spencer (1989, 1) further indicates that "the percentage of the population that is elderly would change from 12.4 percent in 1988 to 13.9 percent in 2010." However, he (1989, 1) notes that "from 2010 to 2030, the number of people 65 and over is projected to increase substantially—from 39.4 million in 2010 to 65.6 million in 2030." Spencer (1989, 1) notes that "nearly 22 percent of the population would be 65 or older in 2030."

In contrast, Spencer (1989, 1) states that "the share of the U.S. population under age 35 may never again be as large as it is now—55 percent." He (1989, 1) then points out that "that percentage is projected to drop to 48 in 2000, 46 in 2010, and 41 in 2030." Spencer (1989, 4) also notes that "one of the most pervasive [population] trends...is the overall aging of the future population." He (1989, 4-7) points to the forecasted increasing age of the American population. Additionally, Silvestri and Lukasiewicz (1989, 51, Table 4) in considering "civilian employment" changes between 1988 and the year 2000 for selected occupations, project "accountants and auditors" as a growing occupation. In fact, they (1989, 60, Table 6) include "accountants and auditors" in a table titled "Occupations with the largest job growth, 1988-2000, moderate alternative projection."

These population projections suggest the possibility that relatively fewer younger individuals will be available to the public accounting profession in the twenty-first century. At present there appears to be a plentiful supply of accounting graduates to meet the needs of public accounting firms (see, for example, Daidone and Knopf 1993, 6, 7, 23, 24) and, as noted later in this paper, over the long term in public accounting it is unlikely that there will be an excess of demand for trained professional personnel over supply of such individuals.[2] However, the issue remains as to the future role of the increasing number of older individuals who have devoted their careers to the public accounting profession.

Prior Literature

These population projections have the potential to affect the age composition of staff in public accounting firms (see, e.g., the discussion in Loeb 1987, 161-162, and Milani et al. 1991, 38) and the personnel policies of public accounting firms (see, e.g., Loeb 1987, 161-162, 164, and Milani et al. 1991, 37, 38, 40, 42). The possible effect of population aging upon accounting is just beginning to be considered in the accounting literature. Loeb (1987) in an editorial in the *Journal of Accounting and Public Policy* discussed the issue of the aging of the American society and commented on a variety of areas in which this phenomenon may affect accounting.[3] A limited portion of that editorial (1987, 161-162, 164) considered personnel issues in public accounting. Milani et al. (1991, 36) discussed the aging of the American society. They (1991) also considered the effect of population aging on public accounting in terms of services offered (pp. 36-38) and personnel policies (pp. 37, 38, 40, 42) relating to the utilization of older professionals by public accounting firms. Some aspects of or associated with the topic of population aging, as it relates to public accounting personnel issues, are mentioned as part of Stigen (n.d.) and Istvan (1991). Huber (1991) considered client services to an older population.

This current paper, while building on existing literature on the topic, expands the discussion of the topic into new areas. Thus, while the discussion of provision of opportunities for older members of the public accounting profession builds on existing literature, the legal, ethical, and public policy discussions presented in the paper move the literature in new directions. The organization of the remainder of the paper is discussed next.

Organization of Remainder of Paper

Legal matters and ethical standards relating to aging and personnel policies in public accounting firms are discussed in the next section of this paper. This is followed by a brief discussion of a limited survey of selected public accounting firms which provided information about some of the issues considered in this paper. The roles of the elderly in American society are discussed in the next section both in general and in relation to the public accounting profession. Next, personnel policies that provide possible opportunities for older members of the accounting profession are discussed in-depth. This discussion is based on ideas in Loeb (1987) and where appropriate related to Stigen (n. d.), Milani et al. (1991), and Istvan (1991). This is followed by a discussion of how providing opportunities for older members of the public accounting profession raises issues relating to "equity" (Moody 1992, 208) between the generations of individuals who are in public accounting. The final section includes a discussion of public accounting profession policy and public policy issues, suggestions for future research, and some conclusions.

LEGAL MATTERS AND ETHICAL STANDARDS RELATING TO AGE AND PUBLIC ACCOUNTING PERSONNEL POLICIES

In the United States the public accounting profession's personnel policies relating to age are subject to a key federal law which is discussed below. Many states also have established laws or some type of legal rules relating to age discrimination and employment (see, for example, the discussion in Eglit 1992e, as updated by Eglit 1992f, 4s-25; Eglit 1994e, 11-95 to 11-101).[4] In contrast, relief under common law for matters relating to age discrimination and employment seems to have been somewhat limited (Eglit 1992a, 15-83 to 15-94).[5] This paper, however, is limited to a brief consideration of the key federal law referred to earlier. In the ethics area, the American Institute of Certified Public Accountants (AICPA) code of ethics considers the issue of age.

Legal Matters

The Age Discrimination in Employment Act of 1967 (known as ADEA) became a federal law in 1967 (see, for example, Bessey and Ananda 1991, 414; Kalet 1990, 2; Eglit and Malin 1992a, 16-4). Bessey and Ananda (1991, 415) note that the law was meant to: (1) encourage the "employment of older persons based on their ability rather than age," (2) "prohibit arbitrary age discrimination in the workplace," and (3) assist "employers and workers" in dealing with the issues related to the affect of aging "on employment." Since its enactment ADEA has been amended several times (see, for example, Bessey and Ananda 1991, 415). Today ADEA generally is applicable "to private employers with 20 or more employees, as well as federal, state, and local governments" (Bessey and Ananda 1991, 415; also see, e.g., Modjeska 1993, Chapter 3, p. 6 (which includes an explanation as to how "20 or more employees" is determined), as well as the discussion on pp. 13-18).

ADEA, in general, proscribes "arbitrary age discrimination" (e.g., Modjeska 1993, Chapter 3, p. 2) against individuals age 40 or older in relation to certain types of "employer practices" including the following: "failing or refusing to hire," "discharging," "compensation," and "terms, conditions, or privileges of employment" (Modjeska 1993, Chapter 3, pp. 21-22, also see pp. 16-20; additionally, see, e.g., Bessey and Ananda 1991, 415-416; Quadagno and Hardy 1991, 470-471; Eglit and Malin 1992a, 16-13 to 16-16, as updated by Eglit 1992c, 2s-91 to 2s-98; Eglit 1994b, 3-2 to 3-12). In general, currently under ADEA a person cannot be required to retire due to age (see, for example, Quadagno and Hardy 1991, 471). There is an exception to the proscription against mandatory retirement if an "employee...is 65 years old and who, for the two-year period immediately before retirement was employed in a bona fide executive or high policymaking position if [the] employee was entitled to an immediate nonforfeitable annual retirement benefit [of] at least $44,000..." (Modjeska 1993, Chapter 3, p. 19; also see, e.g., the discussion in Eglit and Malin 1992a, 16-16.8 to 16-21, as updated by Eglit 1992c, 2s-115 to 2s-118 and Eglit 1994c, 3-54 to 3-66). Also, under ADEA an employer is entitled to certain defenses (see, e.g., Modjeska 1993, Chapter 3, pp. 28-43). For example, Modjeska (1993, Chapter 3, pp. 28-34) notes that one such defense is that age is a factor in qualifying for a particular job.

ADEA applies to employees but does not presently cover a "controlling owner" (*Caruso v. Peat, Marwick, Mitchell & Co.* 1987, 146; see the extended quote from this case below) of a business or professional practice. As will be discussed below, to date the courts have not included partners or their equivalent of public accounting firms who are viewed as owners under the protection of ADEA. However, in general, employees of public accounting firms are protected by ADEA and, as is noted below, in some situations an individual with the title of partner may be viewed as an employee for the purposes of ADEA.

To date, courts have held that a partner in a public accounting firm, who is a partner in the traditional sense of an "owner" of a business (see particularly *Caruso v. Peat, Marwick, Mitchell & Co.* 1987, 148), is not an employee and thus not covered by ADEA. The courts seem to be looking at the substance of the individual's position. The fact that a public accounting firm is organized as a professional corporation owned by "member/shareholders" (*Fountain v. Metcalf, Zima & Company, P.A.* 1991, 1399) does not seem to alter the situation. In *Fountain v. Metcalf, Zima & Company, P.A.* (1991) the court held that where a public accounting firm was organized as a professional corporation with "member/shareholders" (p. 1399), the member/shareholders were in essence partners and were not employees "entitled to sue under ADEA" (p. 1401).

In *Wheeler v. Hurdman* (1987, 277) the court held that for an individual in a public accounting partnership not to be considered an employee, for purposes of being covered by ADEA, the individual in question must be a general partner. In this case a former partner in a national public accounting firm, supported by the U.S. Equal Employment Opportunity Commission, argued that in terms of "economic reality" (1987, 261, also see 258, 262, 268-271) a general partner of a national public accounting firm could be considered an employee. The court accepted the notion that, when viewing the status of a partner in a firm, economic reality should, under some circumstances, be considered (*Wheeler v. Hurdman* 1987, 271). In this case, however, the court held (*Wheeler v. Hurdman* 1987, 276) that the former partner's "participation in profits and losses, exposure to liability, investment in the firm, partial ownership of firm assets, and [the former partner's] voting rights—plus [the former partner's] position under the partnership agreement and partnership laws— clearly placed [the former partner] in a different economic and legal category." Thus, the court held that the former partner was not an employee for purposes of ADEA as well as other "Anti-discrimination Acts" (*Wheeler v. Hurdman* 1987, 277, also see 257, 258).

A similar theme of determining whether an individual was truly a partner in the traditional sense can be found in *Caruso v. Peat, Marwick, Mitchell & Co.* (1987). In that case the court held that "*per se*" the title of "'partner'" does not deny a person protection under ADEA (*Caruso v. Peat, Marwick, Mitchell & Co.* 1987, 148). The court said that "if [the] plaintiff" was a partner "as the term 'partner' is traditionally conceived, [the] plaintiff could not qualify as an employee under the ADEA" (*Caruso v. Peat, Marwick, Mitchell & Co.* 1987, 148). The court did note that "if [the] plaintiff's duties...more closely resembled those of a typical salaried worker, [the] plaintiff may bring an action under the ADEA" (*Caruso v. Peat, Marwick, Mitchell & Co.* 1987, 148).

In *Caruso v. Peat, Marwick, Mitchell & Co.* (1987, 146) the court specifically noted that

> a plaintiff may bring a federal age discrimination action under ADEA only where he is an employee suing his former or current employer.... It is well settled that an individual who has acted as a...controlling owner does not fall within the ADEA definition of "employee," and thus cannot bring an action against the company he once...owned.

The court looked at the following three criteria as a minimum in considering whether a person was a partner or employee in terms of ADEA (*Caruso v. Peat, Marwick, Mitchell & Co.* 1987, 149-150): "(1) the extent of the individual's ability to control and operate his business; (2) the extent to which an individual's compensation is calculated as a percentage of business profits; and (3) the extent of the individual's employment security."[6] Others (e.g., *Simpson v. Ernst & Young* 1994, 1169) have cited these criteria. In *Caruso v. Peat, Marwick, Mitchell & Co.* (1987, 150) the former partner was deemed by the court on all three criteria not to meet the test of partnership status.

In *Simpson v. Ernst & Young* (1994) a U.S. District Court ruled that an individual who had been considered a partner in a Big Six public accounting firm was in terms of ADEA considered to be an employee. In 1989 Ernst & Whinney and Arthur Young & Company merged to form Ernst & Young (*Simpson v. Ernst & Young* 1994, 1162). The resulting firm's "Management Committee" decided that there were too many partners after the merger and the plaintiff in this case was among the partners eventually not retained (*Simpson v. Ernst & Young* 1994, 1164-1165).

The plaintiff, who was in his mid-forties at the time of his discharge, sued under ADEA as well as other laws (*Simpson v. Ernst & Young* 1994, 1162, 1165). While the plaintiff had been the managing partner of Arthur Young & Company's Cincinnati, Ohio office, he did not continue as managing partner of the Cincinnati office of Ernst & Young after the merger (*Simpson v. Ernst & Young* 1994, 1162, 1164). The plaintiff, although he "considered himself to be a partner" of Ernst & Young while with the firm after the merger (*Simpson v. Ernst & Young* 1994, 1165), asked "the Court [to] use an 'economic reality' test rather than traditional legal concepts of determining whether he was a partner or employee" (*Simpson v. Ernst & Young* 1994, 1166). In contrast, "Ernst & Young [argued that...] traditional partnership law" should be used to determine if the plaintiff was a partner (*Simpson v. Ernst & Young* 1994, 1166).

The court in *Simpson v. Ernst & Young,* using a partnership law approach, found (*Simpson v. Ernst & Young* 1994, 1167-1175) that the plaintiff, after the merger, did not have many of the characteristics or rights that a partner (who is an owner) might expect to have under New York state law which the parties had agreed was applicable including, for example, the following: (a) the plaintiff "never established a capital account with" the firm (p. 1170), (b) the plaintiff's remuneration from the firm was not dependent on the firm's profitability (pp. 1170-1171), (c) the plaintiff could not "vote" in partnership admissions or terminations (p. 1172), (d) the plaintiff "had no vote on how firm members were to be compensated" (p. 1172), (e) the plaintiff was not guaranteed the "right to examine the firm's books and records..." (p. 1172), (f) the plaintiff had minimal say in the running of the firm (p. 1172), and (g) the firm's "Management Committee" did not act in a fiduciary capacity towards the plaintiff since, for example, as the firm was terminating partners the firm was also hiring "new" (e.g., p. 1165) professionals (pp. 1165, 1172-1173). In reaching a decision in *Simpson v. Ernst & Young,* the court reviewed other relevant previous cases including *Caruso v. Peat, Marwick, Mitchell & Co.* (1987), *Wheeler v. Hurdman* (1987), and *Fountain v. Metcalf, Zima & Company, P.A.* (1991). The court found in *Simpson v. Ernst & Young* (1994, 1173, 1175) that the plaintiff was an employee (not a partner) and entitled to protection by ADEA. One of the court's other actions was to order a "jury trial on ... age discrimination claims ... " (*Simpson v. Ernst & Young* 1994, 1175). The jury's decision in the trial was a significant monetary judgment in favor of the plaintiff; however, there were at the time indications that the defendant would appeal the decision (see Blum 1994, B1 and Geyelin 1994, B5).

Ethical Standards

The AICPA proscribes age discrimination under Rule 501 of its code of ethics. This rule titled "Acts Discreditable" states that "a member shall not commit an act discreditable to the profession" (AICPA, Continually updated, ET Section 501.01). More specifically, Interpretation 501-2 under this rule states in part that "discrimination based on...age...in hiring, promotion, or salary practices is presumed to constitute an act discreditable...." (AICPA, Continually updated, ET Section 501.03).

MANDATORY PARTNER RETIREMENT IN CERTAIN PUBLIC ACCOUNTING FIRMS

Early in 1994, telephone or personal interviews were conducted by the author with an individual from each of seven different national public accounting firms. These seven individuals were located in local offices of their respective firms and believed by the author to be partners of their respective firms. All seven individuals interviewed indicated that their firm had a mandatory retirement policy for partners. Further, each of the seven individuals noted that this mandatory retirement policy was part of the partnership agreement of their firm. All seven interviewees noted that their firm had a specific mandatory retirement age for partners. The mandatory retirement ages reported by each of the seven interviewees were somewhere in the range of 60 years through and including 65 years of age (exact ages are not given here to preserve anonymity). Further, all seven of the individuals noted that their firm allowed early retirement for partners. The nature of such early retirement varied from firm to firm.

Additionally, an individual from each of four different local/ regional public accounting firms was interviewed by the author by telephone. Individuals interviewed were believed by the author to be partners or the equivalent of partners (in the case of firms that were professional corporations). As noted earlier, under some circumstances, in professional corporations, individuals who are the equivalent of partners are considered partners and are not covered under ADEA. Three of the individuals from local/regional public accounting firms reported that their firms had a mandatory retirement age for partners or the equivalent of partners that was part of their partnership/shareholder agreement, while an individual from the fourth firm noted that the firm did not have mandatory retirement for such individuals. The three firms that were reported to have mandatory ages all had the same mandatory retirement age which was in the mid-sixties (the exact age is not given here to preserve anonymity). Only the three local/regional firms that were reported to have mandatory retirement policies for partners (or their equivalent) were reported to have some possibility for early retirement for such individuals. Again, the nature of the early retirement varied from firm to firm.

As noted above, partners (or their equivalent) of public accounting firms, depending on the circumstances, may not be covered by ADEA. Further, the results of the limited survey reported in this paper suggest that mandatory retirement for partners (or their equivalent) exists in some firms in the public accounting profession.[7] Such mandatory retirement is, if the individuals are truly partners, likely not in conflict with ADEA. It is reasonable to question, whether from both a societal and a public accounting perspective, such mandatory retirement is a good idea given the population patterns mentioned earlier.

THE ELDERLY, THEIR ROLES, AND THE OBLIGATIONS OF THE PUBLIC ACCOUNTING PROFESSION

Riley and Riley (1989, 15) suggest that one of the major issues that our society must currently address is the lack of opportunity for a growing number of elderly individuals to use their abilities. They (1989, 16) point out that "...social science research has clearly demonstrated that the doctrine of inevitable aging decline is a fallacy." These authors (1989, 17) suggest that many of the elderly have reasonable health and successfully live independent lives. Additionally, Knowles (1988, 18) notes that "...there is no evidence that increasing age negatively affects productivity."

Riley and Riley (1989, 17) note that "although death is inevitable, the course of the aging process is not...." They (1989, 17) further assert that "... among older workers, intellectual functioning improves with age if the work situation is challenging and calls for self-direction...." Fyock (1990, 33) states that "increasingly, research proves that chronological age is a poor predictor of physical or mental ability...." Riley and Riley (1989, 18-19) stress the importance of "role opportunities" for the elderly and suggest that inadequate role opportunities can have negative consequences for the elderly (1989, 18).

The issues raised by Riley and Riley (1989) are relevant to the practice of public accounting. For example, a key question that has both ethical and practical implications is whether the public accounting profession has an obligation to its elderly members beyond providing retirement benefits. Does, for example, the public accounting profession have an obligation to provide role opportunities for its elderly members? If for both ethical and practical (such as increasing the number of experienced personnel (see, e.g., the discussion in Milani et al. 1991, 38)) reasons we assume the public accounting profession should meet the needs of its more elderly members for a professional role, then the profession should consider how, in the words of Riley and Riley (1989, 15), to "[adapt] role opportunities and role constraints...at work...." Existing "inadequacies of role opportunities for" (Riley and Riley 1989, 16) elderly members of the public accounting profession can be addressed by what Riley and Riley (1989, 16) call "interventions" which they (1989, 16) note can be used to "reduce these inadequacies." Such interventions have implications for the personnel policies of the firms that comprise the public accounting profession and raise the issue, which is discussed later in this paper, relating to "equity" (Moody 1992, 208) between the generations that compose the public accounting profession.[8]

The use of interventions to provide role opportunities for elderly members of the public accounting profession would be beneficial to that profession, its clients, and society in general. As a result of these role opportunities the public accounting profession would have a larger number of experienced professionals available (see the discussion in Milani et al. 1991, 38). The more recent entrants to the public accounting profession then will have the benefit of an expanded base of experienced professionals from whom they can learn. These experienced professionals can mentor the more recent entrants to the public accounting profession (see Milani et al. 1991, 40).[9] Clients would benefit from continued access to these experienced individuals. Society would benefit from having more elderly individuals employed.

Alternatively, it might be argued that a cost to such interventions is reduced opportunities for younger members of the public accounting profession. For example, a number of individuals interviewed as part of the limited survey discussed earlier cited the need to provide opportunities for younger accountants and/or the need to develop these individuals professionally as the justification of mandatory retirement (also, see the general discussion in Milani et al. 1991, 38).

It is beyond the scope of this paper to attempt to provide a measurement of the benefits and costs of providing role opportunities for the more elderly members of the public accounting profession. However, it is my contention that such interventions should be developed because this is the proper and ethical strategy for the public accounting profession to follow.[10] The next section considers interventions that may provide role opportunities for older members of the public accounting profession.

PERSONNEL POLICIES TO PROVIDE ROLE OPPORTUNITIES FOR OLDER INDIVIDUALS IN PUBLIC ACCOUNTING FIRMS

Loeb (1987, 161-162) suggests that the aging of the American society might have implications for personnel policies of public accounting firms. He (1987, 161) notes that "an emphasis on youth in hiring, long term retention of professional personnel only at the partner or equivalent level, and the permissibility and perhaps emphasis on relatively early retirement may, if such tendencies do exist to any great extent while the general population ages, result in labor shortages in public accounting." Among the possible personnel practices suggested by Loeb (1987, 162) for public accounting firms practicing in an aging society were: (a) "permanent" staff positions and (b) less than full-time "work schedules" for "elderly individuals."[11]

In the middle of the last decade of the twentieth century the specter of excess demand for professional personnel in public accounting now seems unlikely. As noted earlier, at present there seems to be more than an adequate supply of accounting graduates. Further, Barefield (1991, 309) notes "the economics of the CPA business leads the firms to hire experts from many disciplines." He (1991, 309) goes on to note that "the old 'closed' labor market with its heavy emphasis on entry level hiring and promotion from within is shifting to an 'open' labor market with hiring at all levels and less promotion from within."[12] Also, as Copperman and Keast (1983, 2) note "in the long run, labor supply always equals labor demand:... over the long run, the intervening forces of time and price change will act to equalize the amount of labor available to the amount of labor that employers chose to consume." Those authors (1983, 3) note the importance of "time" in such an equation. It seems reasonable, then, to assume that an inadequate supply of professional personnel will not be an immediate problem and may not be a long-term problem for the public accounting profession. Thus, this section of the paper only focuses on personnel interventions that may provide opportunities for older members of the public accounting profession.

Four general types of interventions that may help provide role opportunities for older members of the public accounting profession include: (1) establishing personnel policies that make all ranks in a public accounting firm an attractive career position, (2) defining new and/or expanded roles and job expectations for elderly individuals who elect to remain in public accounting rather than retiring, (3) determining the kinds of benefits that public accounting firms will need to provide to a professional staff that is relatively more elderly, and (4) anticipating and addressing any additional human resource issues that may arise with an increase in the number of elderly professionals working in public accounting firms. These four general types of interventions are to some extent interrelated. Success in utilizing one or more of these interventions may affect the use of the others. These interventions which are discussed in earlier papers are explored in-depth next in this current paper.[13]

All Ranks as Career Positions

Public accounting firms should consider formulating policies relating to matters such as compensation, job tenure, fringe benefits, and client assignments that would make professional positions at all ranks a "career" option (see the discussion in Loeb 1987, 162; also, e.g., Istvan 1991, 47 mentions the concept of "career manager"). Public accounting firms have long had a tradition of hiring and keeping an employee until the individual either becomes a partner or leaves the firm (see Montagna 1974, 23, 49; Loeb 1987, 161-162; Weinstein 1987, 96; Istvan 1991, 47). Weinstein (1987, 96), by quoting a practitioner, suggests that the policy of promotion or turnover is rooted in an ethos of providing opportunities for better qualified individuals. Montagna (1974, 50-51) suggests that by placing staff with a client or a potential client, public accounting firms "strengthen relations with the client, and in some instances they bring a new client to the firm." However, the effectiveness of using former firm employees as vehicles of practice maintenance or development may be diminishing due to increased competitiveness in public accounting in recent years.

At present, the organizational culture of many public accounting firms is not concomitant with the concept of career positions at all ranks. For this particular personnel policy to work, the organizational culture in many public accounting firms would have to change. Further, the nature of the tasks assigned to each career position would have to be given careful consideration. For example, public accounting firms might consider the stress inherent in tasks that are assigned to each position.[14]

Creating career positions at all levels in a public accounting firm provides possible interventions that benefit the public accounting profession and the individuals who work in that profession. One benefit of such a policy would be to assure younger individuals committing to a career in public accounting that positions would be available as such individuals grow older. Alternatively, as discussed later in this paper, this policy has potential to be viewed as a vehicle to limit the opportunities of younger accountants and should be implemented only if it could be accomplished in such a manner that the professional growth of talented younger individuals is not limited.

New and/or Expanded Roles and Job Expectations

Schrank and Waring (1983, 55) note that

work organizations collectively serve as a societal age grade.... Leaving the work organization [—] and concomitantly the labor force—is for some an announcement that middle age is over and that old age has begun.

Thus, in the future instead of encouraging individuals to retire, public accounting firms might seek to define new and/or expanded job roles and expectations for older individuals who may wish to remain with a public accounting firm.

Possible strategies relating to new and/or expanded roles and job expectations for older members of the public accounting profession are considered next. They include serving in a lesser capacity in a firm, working part-time, working as an independent contractor, working as a temporary, and using technology, training, and alternative work sites. These strategies are drawn mostly from literatures in other disciplines and some may seem controversial in terms of the milieu of public accounting. However, given the nature of the issues under consideration, these strategies should at least be considered.

Serving in a Lesser Capacity

As a professional becomes elderly, a public accounting firm might consider having the individual serve in a lesser capacity rather than leave the firm. Schrank and Waring (1989, 122) refer to such a concept as "downward mobility" and note that it is not commonly used "in most organizations." They (1989, 122) suggest that downward mobility is providing an employee with the opportunity to have a job at a smaller salary or lesser rank in an organization instead of leaving the organization. Schrank and Waring (1989, 122) further suggest that downward mobility might be advantageous to an employee in terms of, among other things, an individual's pension and other job "benefits." Copperman and Keast (1983, 59) suggest that downward mobility (they also use other terms, e.g., "downplacement" and "demotion" (p. 59)) is not the prevailing norm in American society. However, Copperman and Keast (1983, 61) suggest that the use of downward mobility in dealing with older employees could have benefits for an employer including allowing opportunities for advancement for younger individuals. Downward mobility, especially from partner to nonpartner status, would be difficult for the public accounting profession to accept. In fact, acceptance of the concept would likely necessitate a change in the organizational culture of many public accounting firms.

In public accounting downward mobility could potentially occur within the partnership, among the staff, or movement from partnership to the staff. Even today some personnel changes in large public accounting firms could be viewed as forms of downward mobility. For example, movement from practice office management (i.e., managing partner) to line partner status or from a national office position to a line position in a local office of a national firm might be viewed by some as downward mobility.[15] A retired partner of a large public accounting firm suggested to the author that some older partners may purposely seek to service less difficult clients or purposely seek out more administrative types of assignments. That individual viewed such strategies as forms of downward mobility. If, as proposed above, all levels in a public accounting firm are attractive professional positions, an older individual in theory might choose to move to a less responsible position rather than to leave a public accounting firm (see the discussion in Milani et al. 1991, 38, 40).

Istvan (1991, 47) suggests that in the past there were two different levels in a number of public accounting firm partnerships which resulted in such firms being a "two-tiered partnership." He (1991, 47) suggests that these firms had a "tier" of partners who were "equity" owners and a "tier" of partners without "equity" (the latter received lower remuneration and usually had less influence in running the firm). Istvan (1991, 47) seems to suggest the current existence of "two-tier partnerships" (p. 47) in a number of public accounting firms. He (1991) also suggests that "some firms have... a two-track system for staff levels" (p. 47) where the tracks are a function of the number of hours worked during a year and how that time is spent (pp. 47-48). As noted earlier (in note 11), Milani et al. (1991, 38) also suggest the existence of the concept of "nonequity" partners. Offering an older equity partner a position as a nonequity partner is an opportunity for downward mobility (see note 11).

Movement from partner to nonpartner status is probably uncommon in public accounting and, in the present environment, would, as noted earlier, likely be difficult. However, the suggested existence, as noted above, of personnel movements that contain some (albeit limited) elements of downward mobility indicates the possibility that over the long-term such a concept is possible in the public accounting profession. Research would be needed before such a policy could be implemented on any large scale basis.

Part-time Work

An alternative role for an older professional is less than full-time work at various levels within a public accounting firm—including at the partnership level. Schedules could be kept flexible to meet the needs of both the firm and the individual.[16] Copperman and Keast

(1983, 38-39) report that generally research has shown favorable reaction to the use of part-time workers. They (1983, 38) note that "...available research indicates that almost any job can successfully be scheduled on a part-time basis." These authors (1983, 39-40) point to the following general advantages of using elderly individuals in part-time positions: (1) the elderly employee usually is as skilled as anyone else, (2) the elderly employee (in comparison to "younger" (p. 39) employees) is more likely to remain with the employer (also see the discussion in Schrank and Waring 1989, 122), and (3) the elderly employee tends to be highly motivated, committed to his or her employer, and as productive as younger employees.

Copperman and Keast (1983, 46) seem to suggest that, when mutually advantageous, employers consider structuring retirement plans as well as policies relating to part-time jobs so that elderly employees can become long-term "permanent part-time" employees. Public accounting firms might want to consider such options. Istvan (1991, 47) suggests that "many Sun Belt firms use retired practitioners" during especially busy periods of time (also, see Milani et al. 1991, 42). While retired practitioners might be hired as part-time employees, they, as discussed more fully below, might also be hired as independent contractors (see the general discussion in Istvan 1991, 44).

Independent Contractors

As suggested above, another strategy for addressing predicted population trends would be for elderly accountants who wish to officially retire, presumably to take advantage of retirement programs and yet work part time in public accounting, to continue in public accounting as independent contractors.[17] Independent contractor/public accounting firm arrangements have existed for decades.[18] Given the predicted population trends, in the future such arrangements may be especially appropriate. An elderly accountant could work for one or more public accounting firms.

Temporaries

Another alternative would be for older accountants to work as "temporaries" for firms that specialize in providing temporary workers (see the general discussion in Stigen n.d., 5 and Fyock 1990,

161-163). Firms that provide temporary accountants for organizations already exist (see, e.g., Rudolph 1986, 59). A 1986 article in *Time,* moreover, suggested that in addition to accountants there was, at that time, a job market for temporary physicians and temporary attorneys as well as temporary professionals in other fields (Rudolph 1986, 59; also see the discussion in Fyock 1990, 161-163).

Technology, Training, and Alternative Work Sites

Czaja and Barr (1989, 129) note that "...new computer and communication technologies have been, and continue to be, rapidly introduced into most occupational settings." These authors (1989, 129) express concern as to how "this technology [may change] the work life of the elderly." They (1989, 129) suggest the inevitable necessity of training elderly workers in such new technology.

Concerted efforts could be made to train elderly as well as younger professionals in public accounting firms in the latest advances in computers and telecommunications as well as other relevant technical developments. This training could include instructing elderly public accounting professionals in how they could effectively use such advances in the practice of public accounting (see the general discussion in Czaja and Barr 1989, 130).[19]

Czaja and Barr (1989, 130) note that "computer technology... makes paid work at home [possible] through...electronic links between office and home and between coworkers." Public accounting firms might consider employing some elderly accountants that are, for health or family reasons, home bound.[20] By the use of computers and telecommunications such elderly individuals could work at home. These individuals could work on matters such as tax preparation or compilations and communicate with both the office and clients through telecommunications.

Benefits Needed

Public accounting firms with professional staffs that become relatively more elderly would need to anticipate and provide appropriate benefits to a relatively older staff. These benefits might be financial (e.g., different types of retirement benefits) and educational (e.g., training programs designed for older staff members; college tuition reimbursement plans for staff wishing to take courses[21]). Health benefits would need to be coordinated with existing government programs or requirements.

Other Possible Human Resource Issues

If the average age of personnel in public accounting firms were to increase substantially additional human resource issues may arise. Two examples of such possible issues which are discussed below, are: (a) providing assurance for all professional staff that opportunities for promotion exist and (b) developing innovative criteria and techniques for evaluating how professional staff and partners perform.

Copperman and Keast (1983, 58) cite studies that suggest that in our society promotions tend to be youth oriented. Such policies would have to be avoided if public accounting firms are to successfully maintain an older work force.

Public accounting firms should use personnel evaluation systems that are fair to older individuals and also do not limit the professional growth of younger individuals. Copperman and Keast (1983, 58) note that "personnel appraisal systems which emphasize qualities often associated with youth—such as aggressiveness or enthusiasm—may be subtly biased against older workers." Avoidance of such an appraisal system in public accounting firms might create pressure between the needs of the older generation and the ambitions of the younger generation. The latter may feel that ignoring their comparative advantages is not equitable. In contrast, the former may feel that an appraisal system, to be equitable, should be age neutral.[22] Such differences in perspectives are discussed in the next section in terms of equity between generations.

The following two additional possible issues were raised by some of the individuals interviewed in the limited survey described earlier: (1) the stressful nature of public accounting and/or (2) the youth oriented culture of public accounting. These issues should be considered when developing interventions for older members of the public accounting profession. For example, public accounting firms should consider ways of changing this youth orientation. Also, in developing interventions consideration should be given to the amount of stress placed on the older professional.[23]

INTERVENTIONS, PUBLIC ACCOUNTING, AND
EQUITY BETWEEN GENERATIONS

Loeb (1987, 163) indicates that there are "potential conflicts that may occur between generations as a society ages." Interventions to increase opportunities for older individuals in public accounting that were mentioned in the previous section of this paper raise the issue of what Moody (1992) refers to as "generational equity" (e.g., p. 208) or "justice between generations" (p. 209). The issue is to what degree one generation should make sacrifices in support of another generation (see Moody 1992, 209). Jonsen (1991, 345-346) notes that

> resentment is a response to deprivation of some value.... Resentment is, then, a sign of presumed injustice. The deprivation results not merely from chance or from the nature of things, but from social arrangements that block access to the enjoyment of the value.

Jonsen (1991, 347-349) suggests that the inability of the elderly to obtain the nonmaterial good "honor" (which he views for the elderly as "being a living part of a society" (p. 347)) can result in "just resentment" (p. 349). Thus, Jonsen (1991, 347-348) in considering resource allocation between generations, suggests that a key issue is "justice" which relates to both "material goods" (p. 348) and nonmaterial goods—particularly honor (also, see Jonsen 1991, 349-350). Moody (1992, 229) suggests that how we act towards the elder generation will "become a precedent for how we ourselves will be treated in turn." This is especially crucial as individuals begin to live longer.

In general, policies in public accounting firms historically favor relatively early retirement (see, e.g., the discussion in Montagna 1974, 58). There has been generally an ethos that emphasizes opportunity for promotion for younger individuals (see, e.g., the general discussion in Weinstein 1987, 94-97). The interventions proposed earlier in this paper that create opportunities for the older individuals to remain useful members of the public accounting profession arguably may be viewed as favoring the more elderly members of the public accounting profession and, at least in the short run, limiting the opportunities for younger members of the public accounting profession (see the general discussion in Milani et al. 1991, 38).

For example, having all ranks in public accounting firms as possible career positions might be viewed as creating a potential conflict between two generations of accountants. Younger accountants may view career positions (e.g., seniors, managers, etc.) as impediments to their own advancement. Further, the use of new and/or expanded roles and job expectations may also create similar conflicts because younger accountants may view the retention of older accountants in any capacity in the public accounting work force as limiting opportunities for the younger generation.

Moody (1992, 208-214) discusses possible problems that may occur when considering "generational equity" (p. 208). He (1992, 208) notes that "*generation...*can mean either a chronological age group...or a historical birth cohort...." Moody (1992, 208) points out that "equity issues involving" the two meanings of generation can be different. He (1992, 209-214) then mentions and comments on a number of possible problems that have been raised about "intercohort equity comparisons" (p. 209). Two examples of these possible problems mentioned by Moody (1992) include: (1) the difficulties in determining cohort "boundaries" (p. 209, also see p. 210) and (2) the difficulties in assessing what is fair or equitable in relation to the various "cohorts" (p. 210, also see p. 211). Thus, there are a number of possible problems relating to the consideration of generational equity that the public accounting profession will need to consider when dealing with the issue of the aging of the population.

PUBLIC ACCOUNTING PROFESSION POLICY AND PUBLIC POLICY IMPLICATIONS, AREAS FOR FUTURE RESEARCH, AND CONCLUSIONS

Public Accounting Profession Policy and Public Policy Implications

The problems that accompany the aging of the American population cannot be avoided by any social institution that is part of our society. Public accounting as a social institution needs to recognize and deal with these problems. Drawing upon the previous section, it follows that a discussion of the issue of equity as it relates to the interests of individuals of different generations in public accounting involves resource allocation. Initially, the public accounting profession should address this resource allocation with some forms of policies that are internal to the profession.[24] Because resource allocation is a central issue, the reality is that any such policy will likely result in a conflict between the generations working in the public accounting profession. Following Jonsen, both material and nonmaterial goods will have to be allocated between the existing generations in public accounting. Nonmaterial goods would include the concept of honor suggested by Jonsen.

Professional associations such as the AICPA and state CPA societies might consider actions to encourage the integration of their elderly members back into the mainstream of the practicing public accounting profession. Efforts could be made on the part of these professional associations to educate their members as to the need to utilize the elderly members of the public accounting profession. These efforts could take place through speeches at conferences by leaders of the public accounting profession, articles in professional journals and newsletters, and continuing education seminars on practice management issues.

As noted earlier, the AICPA code of ethics proscribes discrimination based on an individual's age. In general, most boards of accountancy and most state CPA societies utilize a code of ethics that follow the AICPA's code of ethics (Hermanson et al. 1989, 50). Thus, the AICPA and/or the state CPA societies might publicly indicate that their ethics enforcement bodies would be more proactive in regard to possible age discrimination.

Additionally, because public accounting exists within a larger society, the public policies of the larger society as they relate to the treatment of the elderly will be a key factor affecting public accounting. It is beyond the scope of this paper to recommend particular public policies for the treatment of the elderly. Instead, this paper suggests some possible approaches for providing opportunities and a share of the profession's "goods" to elderly members of the public accounting profession. Suggestions are also given as to how the public accounting profession's associations can encourage this process.

An inadequate response by the public accounting profession to the issues raised in this paper—especially those relating to what in all likelihood will be an increasing number of older members of the public accounting profession—would leave the issue up to public policymakers. For example, pressure may build for changes in ADEA that would eliminate the mandatory retirement of partners in public accounting firms. Boards of accountancy may begin to be more proactive in the area of possible age discrimination. Alternatively, courts may become even stricter in defining who is considered a partner under ADEA. For example, in recent years mergers have resulted in affected national firms increasing in size. These larger sizes and the nature of the partnership agreements may influence courts to view national public accounting firms more like corporations and partners more like employees for purposes of ADEA (see the discussion in *Simpson v. Ernst & Young* 1994). The issue of whether a partner of a large public accounting firm is really an owner of the firm will remain and may depend on the firm and its policies.

The exemption of owners of a professional firm or the owners of a business from ADEA is sensible. It is unlikely that an owner of a professional firm or business would knowingly discriminate against himself or herself. Thus,. a key question for the public accounting profession—especially in relation to large public accounting firms— is, under what circumstances is a partner truly an owner of the firm?[25]

The conflicts and issues discussed in this paper are difficult and yet need to be considered in terms of justly treating each generation. Historically, as suggested earlier, the emphasis in public accounting appears to have been on opportunities for the younger generation. As our society becomes older it seems reasonable, practical, and just to provide more role opportunities for the older generation. Careful tradeoffs may be needed so that the needs of both generations can be considered and, at the same time, the public interest missions of the public accounting profession can be accomplished without further involvement of public policymakers.[26]

Areas for Future Research

The issue of population aging and personnel policies of public accounting firms presents a number of opportunities for future research. For example, research could be directed at the feasibility of the interventions suggested in this paper. This would include the strategies suggested relating to new and/or expanded roles and job expectations for older members of the accounting profession (downward mobility; part-time work; expanded use of older accountants as independent contractors or temporaries; use of technology, training, and alternative work sites). Research could examine the economic feasibility and practicality of creating career positions at all ranks in public accounting firms. Research could also consider how the various professional associations that serve the public accounting profession could deal with the changing age demographics. Also, research could consider how the public accounting profession in nations other than the United States deals with demographic issues relating to age.[27] Finally, research could consider how the courts may view, in terms of ADEA, the status of partners of the large public accounting firms which have become limited liability partnerships (LLPs).[28]

Conclusions

The projections of a longer living American population will have personnel implications for public accounting firms. These firms should consider the needs and problems of their older members. As noted earlier, there are a number of personnel issues that public accounting firms may wish to consider. Providing role opportunities for such older individuals should be a priority issue for the public accounting profession. However, while considering the needs of such individuals, care should be given to maintain equity between the generations of those practicing public accounting.

ACKNOWLEDGMENTS

I am indebted to a number of individuals for their comments and suggestions including Mark Mitschow, Kwok Leung, Nile Webb, C. Michael Nath, Stephen Liedtka, Augustine Duru, Donal Byard, Cindy Loeb, Dan Ostas, and Felicia LeClere. Cathy Ventrell-Monees, Esquire, Manager, Worker Equity Section, American Association of Retired Persons provided many suggestions relating to legal issues. Finally, I am especially indebted to Larry Parker and two anonymous referees of *Research in Accounting Regulation* for their thoughtful suggestions. Any errors are my own.

NOTES

1. See, for example, Pifer and Bronte (1986b, especially p. 4), Taeuber (1983, 1, 25), Copperman and Keast (1983, 12, 15), and Loeb (1987, 157-158). Also, see the projections in Sarkissian (1989, 44, 46) and Milani et al. (1991, 36). Further, see the discussion in Nelson (1989, 46). Also, see Palmer and Gould (1986, 374), Fyock (1990, 1, 3, 4), Huber (1991, 46), and Istvan (1991, 47).

2. However, Daidone and Knopf (1993) do not report on the ages of accounting graduates.

3. Many of the citations for Loeb (1987) come from various papers in Pifer and Bronte (1986a). See Loeb (1987, 158, note 2).

4. Possible sources of protection from age discrimination and employment at the state level (depending on the state (see Eglit 1992e, 20-1 to 20-11; Eglit 1994e, 11-95 to 11-101)) may be found in (1) state laws specifically relating to this issue (Eglit 1992e, 20-1 to 20-2; Eglit 1994e, 11-95 to 11-96), (2) state laws other than a law specifically on this issue (Eglit 1992e, 20-2; Eglit 1994e, 11-96), (3) "executive orders" (Eglit 1992e, 20-2; Eglit 1994e, 11-96), and (4) the constitution of the state (e.g., Eglit 1992e, 20-2, as updated by Eglit 1992f, 4s-25; Eglit 1994a, 2-21). An individual may bring legal action relating to age discrimination and employment under ADEA, state law, or both (see, for example, Eglit and Malin 1992b, 17-45, as updated by Eglit 1992d, 3s-73; Modjeska 1993, Chapter 3, p. 3; Eglit 1994a, 2-22 to 2-26; Eglit 1994c, 6-247 to 6-266). In fact, certain state laws may be more protective when compared to ADEA (Eglit and Malin 1992b, 17-45; Eglit 1994a, 2-22). However, a state law cannot cause ADEA to be invalid (see, for example, Eglit and Malin 1992b, 17-47, as updated by Eglit 1992d, 3s-74 to 3s-75; Modjeska 1993, Chapter 3, p. 3; Eglit 1994a, 2-24 to 2-25). See, for example, Eglit and Malin (1992b, 17-6 to 17-7, 17-17 to 17-20, 17-45 to 17-71, 17-73 to 17-77, as updated by Eglit 1992d, 3s-2, 3s-17 to 3s-21, 3s-73 to 3s-97, 3s-100 to 3s-102), Eglit (1994a, 2-22 to 2-26), and Eglit (1994c, 6-4 to 6-7, 6-152 to 6-166, 6-242 to 6-245, 6-247 to 6-266) for a discussion of the relationship of state law and ADEA. Finally, Eglit (1994e, 11-95) indicates that certain local governments have laws relating to age discrimination and employment.

5. Eglit (1992a, 15-94) suggests that, in general, while "nonstatutorily based" approaches to age discrimination and employment have not been too successful, common law approaches may have some future potential. The possible potential of a common law approach in certain circumstances is supported by Eglit (1992b, 2s-79 to 2s-88) and Eglit (1994d, 10-97 to 10-123).

6. Interestingly, the same court in further matters relating to this case later noted that "a partner is generally considered a permanent employee, who cannot be fired or released except in extraordinary circumstances" (*Caruso v. Peat, Marwick, Mitchell & Co.* (1989, 222).

7. I am indebted to an anonymous reviewer for raising issues that resulted in the limited survey discussed in this current paper being conducted. When necessary, additional contact was made with several individuals interviewed to clarify answers. The existence of mandatory retirement in public accounting is also indicated by Milani et al. (1991, 38) who suggest the reexamination of "the mandatory retirement policies found at many public accounting firms." Also, see the other comments on mandatory retirement in public accounting in Milani et al. (1991, 38) including the quote from that source in note 11 of this current paper. Also, see, e.g., Cowan (1994).

8. As suggested above, the topic of the public accounting profession and its obligations to its elderly members has possible ethical implications. Loeb (1987, 160, 164-165) notes additional areas in which population aging may raise ethical issues for accountants. He (p. 160) raises the question as to whether "accountants [should use] their skills to gather and/or analyze data that possibly could be used to deny or reduce health services to elderly individuals." Also, he raises the issue "as to whether the accounting occupation and/or accounting research should assist in the development and/or refinement of techniques that may involve the allocation or restriction of health care resources to the elderly" (p. 160). Also see Loeb (1987, 164-165).

9. I am indebted to an anonymous reviewer for emphasizing to me the importance of this point. Also, see the general discussion in Alter (1991, 52, 55).

10. I am indebted to an anonymous reviewer for suggesting the idea on which the argument is based.

11. Also, see the discussion in Hooks (1990), Stigen (n.d., especially pp. 6-9), and Milani et al. (1991, 38, 40, 42). Milani et al. (1991, 38) note that "a number of public accounting firms now offer nonequity partner alternatives for desirable specialists past the mandatory retirement age."

12. I am indebted to an anonymous reviewer for pointing out this source and the two sentences quoted here.

13. These four general interventions include explicitly or implicitly the two personnel practices suggested by Loeb (1987, 162) that were mentioned earlier. Also, see Loeb (1987, 163, 164) for a brief discussion of training which is noted later, in this current paper, as a benefit that may be needed by older public accountants. Also, interventions 2, 3, and 4 are considered or can be related to in Milani et al. (1991, 38, 40, 42). For interventions relating to gender issues see, for example, Alter (1991) and AICPA (1993).

14. At present some major public accounting firms are talking publicly about the possibility of reducing staff turnover and retaining individuals longer in staff positions. Also, see the general discussion in Weinstein (1987, 96), Stigen (n.d., 6), and Alter (1991, 55). Also, the comment relating to stress is based on a suggestion by an anonymous reviewer.

15. See note 11 in this current paper. Also, see the discussion in Milani et al. (1991, 38).

16. See Stigen (n.d., 8-9) and Fyock (1990, 154-161). Also, see the discussion in Milani et al. (1991, 38, 40) and Copperman and Keast (1983, 6, 34-50). Also, generally see Taeuber (1983, 23), Alter (1991, 51, 52, 55), and AICPA (1993).

17. Stigen (n.d., p. 8) mentions the possibility of engaging accounting professionals on such a basis. Also, see the general discussion in Fyock (1990, 164-165).

18. I can recall such arrangements existing in the 1960s.

19. See Fyock (1990, 98-121) for a discussion of training issues relating to the elderly. See the discussion in Milani et al. (1991, 38, 40), Copperman and Keast (1983, 54-57), and Schrank and Waring (1989, 120-121). Also, see Greenberg (1988, 27) for a brief discussion of the training that one company used when employing older workers.

20. Nelson (1989, 52) indicates that "'Flex place' (working in the home) will need to be increased to allow two wage-earner households to be more productive." Milani et al. (1991, 42) also note the possibilities of using computers to work at home. Also, see the general discussion in Stigen (n.d., 4), Istvan (1991, 4), and Alter (1991, 55).

21. See Schrank and Waring (1989, 120-121). See Strategic Planning Committee (1988, 43) which mentions the idea of "portable benefits." Also, see Loeb (1987, 163), Fyock (1990, 128-143), Milani et al. (1991, 38, 40, 42), and Copperman and Keast (1983, 6, 33, 34, 68-98).

22. See Schrank and Waring (1989, 120) for a discussion of the evaluation of older workers. Also, see Fyock (1990, 173-174).

23. I am indebted to an anonymous reviewer for suggesting the idea on which this argument is based.

24. This is in contrast to policies set outside the profession by public policymakers. I am indebted to an anonymous reviewer for stressing the importance of the discussion in this section of the paper. Also, see the general discussion in Palmer and Gould (1986, 374).

25. I am indebted to Dan Ostas for suggesting the idea on which the argument is based

26. I am indebted to an anonymous reviewer for suggesting the consideration of public and professional policy issues. See note 24.

27. I am indebted to an anonymous reviewer for suggesting the idea on which this section and particularly the comment on other nations is based.

28. See, for example, the general discussion concerning LLPs and national public accounting firms and LLPs in "LLP Becomes the Closing of Choice" (1994, 8).

REFERENCES

American Institute of Certified Public Accountants (AICPA). Continually updated. *AICPA Professional Standards, Vol. 2.* New York: American Institute of Certified Public Accountants.

_____. 1993. *AICPA Resource Clearinghouse on Women and Family Issues in the Accounting Workplace.* [*New York*]: American Institute of Certified Public Accountants.

Alter, J. 1991. Retaining women CPAs. *Journal of Accountancy* (May): 50-52, 55.

Barefield, R.M. 1991. A critical view of the AECC and the converging forces of change. *Issues in Accounting Education* (Fall): 305-312.

Bessey, B.L., and S.M. Ananda. 1991. Age discrimination in employment: An interdisciplinary review of the ADEA. *Research on Aging* (December): 413-457.

Blum, A. 1994. Fired partner wins case on age bias. *The National Law Journal,* May 30, p. B1.

Caruso v. Peat, Marwick, Mitchell & Co. 664 F. Supp. 144 (S.D.N.Y. 1987).

_____. 717 F. Supp. 218 (S.D.N.Y. 1989).

Copperman, L.F., and F.D. Keast. 1983. *Adjusting to an Older Work Force.* New York: Van Nostrand Reinhold.

Cowan, A.L. 1994. Unmourned departure at Coopers. *The New York Times,* January 17, pp. D1, D8.

Czaja, S.J., and R.A. Barr. 1989. Technology and the everyday life of older adults. *The Annuals of the American Academy of Political and Social Science* (May): 127-137.

Daidone, J., and (in conjunction with) L.W. Knopf. 1993. *The Supply of Accounting Graduates and the Demand for Public Accounting Recruits—1993.* New York: American Institute of Certified Public Accountants.

Eglit, H.C. 1992a. Employment discrimination: Constitutional and other nonstatutory parameters. In *Age Discrimination, Volume 2,* H.C. Eglit (with M.H. Malin), Chapter 15. Colorado Springs, CO: Shepard's/McGraw-Hill, Inc. For this edition the earliest copyright date noted is 1981.

_____. 1992b. Employment discrimination: Constitutional and other nonstatutory parameters. In *Age Discrimination: 1992 Cumulative Supplement to Volume 2. Current through all statutes, cases, and regulations published as of November 4, 1991*, Chapter 15. Which is in *Age Discrimination, Volume 2*, H.C. Eglit (with M.H. Malin). Colorado Springs, CO: Shepard's/McGraw-Hill, Inc. For this edition the earliest copyright date noted is 1981.

_____. 1992c. Employment discrimination: The Age Discrimination in Employment Act—protections, prohibitions and exceptions. In *Age Discrimination: 1992 Cumulative Supplement to Volume 2. Current through all statutes, cases, and regulations published as of November 4, 1991*, Chapter 16. Which is in *Age Discrimination, Volume 2*, H.C. Eglit (with M.H. Malin). Colorado Springs, CO: Shepard's/McGraw-Hill, Inc. For this edition the earliest copyright date noted is 1981.

_____. 1992d. Employment discrimination: The Age Discrimination in Employment Act—Enforcement. In *Age Discrimination: 1992 Cumulative Supplement to Volume 3. Current through all statutes, cases, and regulations published as of August 5, 1991*, Chapter 17. Which is in *Age Discrimination, Volume 3*, H.C. Eglit (with M.H. Malin). Colorado Springs, CO: Shepard's/McGraw-Hill, Inc. For this edition the earliest copyright date noted is 1981.

_____. 1992e. Employment discrimination: State regulation. In *Age Discrimination, Volume 4*, H.C. Eglit (with M.H. Malin), Chapter 20. Colorado Springs, CO: Shepard's/McGraw-Hill, Inc. For this edition the earliest copyright date noted is 1981.

_____. 1992f. Employment discrimination: State regulation. In *Age Discrimination: 1992 Cumulative Supplement to Volume 4. Current through all statutes, cases, and regulations published as of August 5, 1991*, Chapter 20. Which is in *Age Discrimination, Volume 4*, H.C. Eglit (with M.H. Malin). Colorado Springs, CO: Shepard's/McGraw-Hill, Inc. For this edition the earliest copyright date noted is 1981.

_____. 1994a. The Age Discrimination in Employment Act—Introduction. In *Age Discrimination, Second Edition, Volume 1*, H.C. Eglit, Chapter 2. Colorado Springs, CO: Shepard's/McGraw-Hill, Inc. The copyright date for this edition is 1993.

_____. 1994b. The Age Discrimination in Employment Act—Who is protected and which entities must comply with its requirements. In *Age Discrimination, Second Edition, Volume 1*, H.C. Eglit, Chapter 3. Colorado Springs, CO: Shepard's/McGraw-Hill, Inc. The copyright date for this edition is 1993.

_____. 1994c. Enforcement of the Age Discrimination in Employment Act in the nonfederal sector. In *Age Discrimination, Second Edition, Volume 1*, H.C. Eglit, Chapter 6. Colorado Springs, CO: Shepard's/McGraw-Hill, Inc. The copyright date for this edition is 1993.

_____. 1994d. Employment discrimination: Constitutional and other nonstatutory parameters at the federal and state levels. In *Age Discrimination, Second Edition, Volume 3*, H.C. Eglit, Chapter 10. Colorado Springs, CO: Shepard's/McGraw-Hill, Inc. The copyright date for this edition is 1993.

————. 1994e. Other bais to age discrimination in employment: Federal and state statutes, executive orders, and regulations. In *Age Discrimination, Second Edition, Volume 3,* H.C. Eglit, Chapter 11. Colorado Springs, CO: Shepard's/McGraw-Hill, Inc. The copyright date for this edition is 1993.

Eglit, H.C., and M.H. Malin. 1992a. Employment discrimination: The Age Discrimination in Employment Act—protections, prohibitions and exceptions. In *Age Discrimination, Volume 2,* H.C. Eglit (with M.H. Malin), Chapter 16. Colorado Springs, CO: Shepard's/McGraw-Hill, Inc. For this edition the earliest copyright date noted is 1981.

————. 1992b. Employment discrimination: The Age Discrimination in Employment Act—enforcement. In *Age Discrimination, Volume 3,* H.C. Eglit (with M.H. Malin), Chapter 17. Colorado Springs, CO: Shepard's/McGraw-Hill, Inc. For this edition the earliest copyright date noted is 1981.

Fountian v. Metcalf, Zima & Company, P.A. 925 F. 2d 1398 (11th Cir. 1991).

Fyock, C.D. 1990. *America's Work Force is Coming of Age: What Every Business Needs to Know to Recruit, Train, Manage, and Retain an Aging Work Force.* Lexington, MA: Lexington Books.

Geyelin, M. (M.A. Jacobs contributed to this article). 1994. Legal beat: Law notes.... *The Wall Street Journal,* May 17, p. B5.

Greenberg, B.R. 1988. A comprehensive company approach. In *Employing Older Americans: Opportunities and Constraints.* Summary of a Symposium, September 21-23, 1987, Wingspread Conference Center, Racine, WI. Research Report No. 916, ed. H. Axel, 23-27. New York: The Conference Board.

Hermanson, R.H., J.R. Strawser, and R.H. Strawser. 1989. *Auditing Theory and Practice,* 5th ed. Homewood, IL: Irwin.

Hooks, K.L. 1990. Let's give alternative work schedules a chance. *Journal of Accountancy* (July): 81, 82, 84, 86.

Huber, E. 1991. Protecting the assets of elderly clients. *The CPA Journal* (May): 46-48, 50, 52, 53.

Istvan, D.F. 1991. Coming full circle in practice management. *Journal of Acocuntancy* (May): 42-44, 47, 48.

Jonsen, A.R. 1991. Resentment and the rights of the elderly. In *Aging and Ethics: Philosophical Problems in Gerontology,* ed. N.S. Jecker, 341-352. Clifton, NJ: Humana Press.

Kalet, J.E. 1990. *Age Discrimination in Employment Law,* 2nd ed. Washington, DC: The Bureau of National Affairs.

Knowles, D.E. 1988. Dispelling myths about older workers. In *Employing Older Americans: Opportunities and Constraints.* Summary of a Symposium, September 21-23, 1987, Wingspread Conference Center, Racine, WI. Research Report No. 916, ed. H. Axel, 16-18. New York: The Conference Board.

LLP becomes the closing of choice. 1994. *The CPA Journal* (October): 8.

Loeb, S.E. 1987. Population aging: Some accounting considerations. *Journal of Accounting and Public Policy* (Fall): 157-167.

Milani, K., J. Barsanti, and M. Egan. 1991. The graying of America. *Journal of Accountancy* (February): 36-38, 40, 42.

Modjeska, A.C. 1993. *Employment Discrimination Law,* 3rd ed. Deerfield, IL: Clark Boardman Callaghan.

Montagna, P.D. 1974. *Certified Public Accounting: A Sociological View of a Profession in Change.* Houston: Scholars Book Co.

Moody, H.R. 1992. *Ethics in an Aging Society.* Baltimore: The Johns Hopkins University Press.

Nelson, A.T. 1989. The human resource dilemma in accounting. *Journal of Acocuntancy* (August): 46-48, 50, 52.

Palmer, J.L., and S.G. Gould. 1986. Economic consequences of population aging. In *Our Aging Society: Paradox and Promise,* eds. A. Pifer and L. Bronte, 367-390. New York: W.W. Norton.

Pifer, A., and L. Bronte, eds. 1986a. *Our Aging Society: Paradox and Promise.* New York: W.W. Norton.

Pifer, A. and L. Bronte. 1986b. Introduction: Squaring the pyramid. In *Our Aging Society: Paradox and Promise,* eds. A. Pifer and L. Bronte, 3-13. New York: W.W. Norton.

Quadagno, J.S., and M. Hardy. 1991. Regulating retirement through the Age Discrimination in Employment Act. *Research on Aging* (December): 470-475.

Riley, M.W., and J.W. Riley, Jr. 1989. The lives of older people and changing social roles. *The Annuals of the American Academy of Political and Social Science* (May): 14-28.

Rudolph, B. (reported by C. Garcia, with other bureaus). 1986. Help wanted: Professional temps in demand. *Time,* October 6, p. 59.

Sarkissian, R.V. 1989. Retail trends in the 1990s. *Journal of Accountancy* (December): 44-46, 48, 50, 53, 55.

Schrank, H.T., and J.M. Waring. 1983. Aging and work organizations. In *Aging in Society: Selected Reviews of Recent Research,* eds. M.W. Riley, B.B. Hess, and K. Bond, 53-69. Hillsdale, NJ: Lawrence Erlbaum.

_____. 1989. Older workers: Ambivalence and interventions. *The Annuals of the American Academy of Political and Social Science* (May): 113-126.

Silvestri, G., and J. Lukasiewicz. 1989. Projections of occupational employment, 1988-2000. *Monthly Labor Review* (November): 42-65.

Simpson v. Ernst & Young. 64 FEP Cases 1161 (1994).

Spencer, G. 1989. *Projections of the Population of the United States, by Age, Sex, and Race: 1988 to 2080.* U.S. Bureau of the Census, Current Population Reports, Series P-25, No. 1018. Washington, DC: U.S. Government Printing Office.

Stigen, R.A. n.d. The CPA firm of the future. *Accounting Today: CPE Program,* Lesson 4.

Strategic Planning Committee. 1988. *Strategic Thrusts for the Future,* 1st ed.: *Report of the Strategic Planning Committee.* [*New York*]: American Institute of Certified Public Accountants, October.

Taeuber, C.M. 1983. *America in Transition: An Aging Society.* U.S. Bureau of the Census, Current Population Reports, Series P-23, No. 128. Washington, DC: U.S. Government Printing Office.

Weinstein, G.W. 1987. *The Bottom Line: Inside Accounting Today.* New York: NAL Books.

Wheeler v. Hurdman. 825 F. 2nd 257 (10th Cir. 1987).

PART III

RESEARCH REPORTS

AIDING AND ABETTING
AFTER CENTRAL
BANK OF DENVER

Mark A. Segal

ABSTRACT

Considerable litigation has been brought against accountants for allegedly aiding and abetting Section 10(b)(5) violations. Recently the Supreme Court ruled in the case of *Central Bank* that no implied private cause of action exists for aiding and abetting Section 10(b) violations. The ruling and its rationale appear to signify a turning point in the development of accountant liability. In this paper this important case and its implications are examined. Understanding *Central Bank* and its ramifications is important to determining the standards of conduct to which accountants will be subject, and to enacting legislation which will meet the interests of both the accounting profession and the investing public.

Research in Accounting Regulation, Volume 9, pages 153-162.
Copyright © 1995 by JAI Press Inc.
All rights of reproduction in any form reserved.
ISBN: 1-55938-883-8

INTRODUCTION

In 1966 judicial recognition was accorded for the first time to the existence of an implied private cause of action under Section 10(b) of the Securities Exchange Act of 1934 against persons who aid and abet violations of Section 10(b) or 10(b)(5) (*Brennan* 1966). Since 1966 circuit courts of appeal have consistently recognized the existence of this theory of liability (Wager and Failla 1994). In recent years allegations of aiding and abetting have been commonly raised against accountants (*Farlow* 1991; *Robin* 1990; *Roberts* 1988; *Zoelsch* 1987; *Rudolph* 1986). Thus, it was of significance to the accounting profession when on April 19, 1994, in *Central Bank of Denver, N.A. v. First Interstate Bank of Denver, N.A.*, the Supreme Court rejected this long recognized implied cause of action.

The ramifications of the *Central Bank* decision extend beyond its application to cases involving private actions for aiding and abetting claims. When viewed in conjunction with the Supreme Court's 1993 decision in *Reves*, *Central Bank* reflects the Supreme Court's assertive overturning of what, based upon lower court precedent, had appeared to be recognized law. In reaching these decisions the Court relied on strict statutory construction. Maintenance of this approach by courts will not only affect a substantial number of pending and ongoing lawsuits of relevance to accountants, but other securities actions involving claims not based on the express wording of the law. For example, a strict construction of securities statutes raises questions as to the SEC's ability to bring enforcement actions based on aiding and abetting, as well as whether recklessness will support a claim of primary 10(b)(5) liability, and whether notions of respondeat superior and agency apply in certain contexts of the securities law.

This study examines the *Central Bank* decision and its implications for accountant liability. Examining *Central Bank* is important to understanding the legal standards to which accountants will be subject, and to determining the standards which should exist and can be feasibly enacted.

AIDING AND ABETTING

Aiding and abetting is a form of secondary liability as it is imposed on parties due to their relationship with a primary wrongdoer rather than their actual violation of the relevant statute (Fischel 1981; Kadish 1985). In order to maximize prospects of recovery, allegations of aiding and abetting have been frequently raised against accountants and other secondary parties. Pursuant to the theory of aiding and abetting, a party can be held subject to joint and several liability with the principal perpetrator of a 10(b)(5) violation where:

1. there existed a primary 10(b) or 10(b)(5) violation;
2. the party had knowledge of the primary violation (or was reckless with respect to the primary violation); and
3. the party substantially assisted commission of the primary violation. (*Central Bank 1994*).

According to former SEC Chairperson David Ruder (1994), vulnerability to being held liable for the entire amount of damages resulting from a 10(b) or 10(b)(5) violation has frequently resulted in secondary defendants (e.g., accountants and attorneys), entering into settlements of large class actions suits for "significant sums." Where the primary violator faces financial difficulty due to the unlikelihood of receiving contribution, the pressure to enter into settlements is increased.

Uncertainty over how expansively courts would apply the elements of aiding and abetting added to concern over how these type actions would be resolved in litigation. Application of an expansive approach is reflected in the case of *American Continental Corporation/ Lincoln Savings and Loan Litigation* (1992). The case involved an action against the accounting firm of Touche Ross with respect to the fraudulent acts of John Keating and Lincoln Savings and Loan. Touche had been retained by Lincoln during the course of a public offering. In an 8K filed with the SEC, Touche was noted to have been engaged and to have indicated that the treatment of certain transactions by Lincoln conformed to that prescribed in the accounting literature. Touche did not complete an audit or issue an audit report, nor was it identified in any other statement or document. Nevertheless, the Court denied Touche's motion for summary judgment. In the Court's opinion Touche could potentially be found to have knowledge and substantially assisted the commission of the fraud, and could not "enjoy pecuniary benefit of an engagement, while protecting itself by postponing the issuance of an audit opinion, or making highly qualified representation to the public or regulators."

Even though aiding and abetting has been recognized by the Circuit Courts prior to Central Bank the Supreme Court had neither expressly accepted nor rejected recognition of aiding and abetting liability for 10(b)(5) violations.

CENTRAL BANK

Despite its importance to the accounting profession, *Central Bank* did not per se involve accountants. The case concerned the issuance of $26 million in bonds to finance public improvements for a residential commercial development. Central Bank served as indenture trustee for the bond issue. Subsequently, the bonds went into default and investors sued Central Bank for aiding and abetting a 10(b)(5) violation. At the core of the suit was the assertion that Central Bank should have been aware of the inadequacies of an appraisal of the collateral for the bonds. By not reporting the inadequacy the bank was alleged to have assisted in the commission of a 10(b)(5) violation.

Initially, the case was heard by the District Court for Colorado. The District Court granted summary judgment in favor of Central Bank. The decision was reversed by the 10th Circuit. The 10th Circuit held that Central Bank could potentially be held responsible for aiding and abetting if it were established that:

1. a primary violation of Section 10(b)(5) had been alleged;
2. Central Bank had been reckless; and
3. Central Bank had provided substantial assistance to the primary violator in committing the primary violation.

Writ of certiorari was filed with the Supreme Court with regard to whether recklessness was sufficient to support a finding of liability for aiding and abetting. Rather than confront the narrow issue presented to it, the Court analyzed whether a private action for aiding and abetting under Section 10(b) should be recognized at all. Twice before the Supreme Court had been presented with an opportunity to deal with this issue and had refrained from doing so (*Ernst and Ernst v. Hochfelder* 1976; *Herman and MacClean v. Huddleston* 1983). Now, despite the consistent recognition of aiding and abetting by lower federal courts, and the validity of the concept of aiding and abetting liability not having been challenged by the parties before it, the Supreme Court raised the issue on its own initiative.

In a 5-4 decision, the Court held that no private cause of action for aiding and abetting could be implied under Section 10(b). Writing for the majority, Justice Kennedy based the refusal to recognize aiding and abetting on strict statutory construction. According to the Court the expression "direct or indirect" found in Section 10(b) did not pertain to aiding and abetting. Had allowance of aiding and abetting been intended, the Court reasoned, Congress would have utilized appropriate language. The use of such language where intended with respect to aiding and abetting actions is evident in other federal statutes, for example, Internal Revenue Code Section 6701 and Commodities Exchange Act Section 25(a). In addition, tacit approval of aiding and abetting could not be inferred from Congress not having enacted legislation in light of lower federal court action.

The Court asserted that its decision was based on statutory construction and not policy grounds. In the Court's words, "the issue, however, is not whether imposing private civil liability on aiders and abettors is good policy but whether aiding and abetting is covered by statute." Nevertheless, the Court did mention policy matters, and did so in a manner reflecting a concern with an overly expansive standard of liability (Seligman 1994). While recognizing that good arguments could be made both for and against aiding and abetting liability, the Court noted that adoption of too broad a standard of liability would expose parties to vexatious lawsuits, and substantial costs in mounting a defense and satisfying settlements and court decisions. With respect to economic implications the Court noted:

> This uncertainty and excessive litigation can have ripple effects. For example, newer and smaller companies may find it difficult to obtain advice from professionals. A professional may fear that a newer or smaller company may not survive and that business failure would generate securities litigation against the professional, among others. In addition, the increased costs incurred by professionals because of the litigation and settlement costs under Section 10(b)(5) may be passed on to their client companies and in turn incurred by the companies investors, the intended beneficiaries of the statute.

IMPLICATIONS OF CENTRAL BANK

The Short Term

The *Central Bank* decision has had a significant effect on private litigation. In light of *Central Bank* several private actions for aiding and abetting against accounting firms have been dismissed or summary judgment granted to the accounting firm (*Vosgerichian* 1994; *Cortec Industries* 1994; *In Re: Medeva Securities Litigation* 1994; *Teague* 1994; *Hayes* 1994; *In Re: Kendall Square Research Securities Litigation* 1994). Big six firms benefiting from these decisions have included Deloitte, Haskins and Sells, Arthur Andersen, Price-Waterhouse, and Ernst and Young. In response to *Central Bank* certain plaintiffs have sought to amend allegations of aiding and abetting to assertions of a primary violation. These attempts have not always been successful (*Vosgerichian* 1994).

The pursuit of recovery based on assertions of a primary violation, in lieu of aiding and abetting, operates to the benefit of the accountants. Reliance on a claim of a primary violation rather than a claim of aiding and abetting should make it more difficult to succeed in litigation against parties such as accountants, and as a consequence alleviate pressure placed on these parties to enter into settlements In addition, it is expected that certain actions that previously would have been raised as claims of aiding and abetting will not be asserted as primary violations due to obstacles faced in avoiding dismissal of the claim, and successfully establishing a primary violation.

As expressed by the dissent in *Central Bank* the majority opinion can be read as preventing the SEC from bringing Section 10(b)(5) aiding and abetting enforcement actions. Application of *Central Bank* in this manner casts doubt on long-standing SEC practices (Securities Act Release No. 5088 1970). Shortly after *Central Bank* both the SEC's General Counsel Simon Lorne and Chairperson Arthur Levitt indicated their belief that the decision should not apply to SEC enforcement actions ("SEC Discussing" April 29, 1994; Hearings 1994). SEC Chairperson Levitt indicated that the SEC might pursue litigation over *Central Bank* in selected cases, although it would not devote substantial resources to litigating the issue (Hearings 1994).

The SEC's concern with the potential application of *Central Bank* to aiding and abetting enforcement actions is evident in the SEC's attempts to replead some 26 enforcement actions regarding aiding and abetting (Bureau of National Affairs 1994a). Whether the SEC will be able to successfully replead these cases is uncertain. In the recent case of *Militano* the SEC was denied the ability to amend a complaint in light of *Central Bank* due to such amendment being found unduly prejudicial in that it would increase the financial exposure of the defendants, and increase the need for further discovery (*Militano* 1994).

The Long Term

Shortly after *Central Bank* events transpired raising questions as to whether the decision would be legislatively overturned, or judicial decisions provide an exception for SEC enforcement actions. In this regard Senator Metzenbaum proposed legislation which would reverse *Central Bank* (S.2306) and the SEC indicated that it might challenge application of the decision to SEC enforcement actions. As time has progressed, however, the long-term significance of *Central Bank* has become more assured. Senator Metzenbaum has retired; the SEC has indicated that it will not pursue legislation to reform judicial decisions and has adopted a policy of repleading its aiding and abetting actions (Bureau of National Affairs 1994a, 1994b). In addition, the assumption of control of both the House of Representatives and the Senate by the Republican Party has decreased the likelihood that legislation reversing *Central Bank* will be enacted unless as part of a more comprehensive bill containing provisions also favorable to defendants in securities litigation.

Provisions of bills introduced in the 103rd Congress by Representative Tauzin (H.R. 417) and Senators Dodd and Domenici (S. 1976) may serve as a starting point for discussion of compromise legislation. These bills favor the discouragement of abusive or frivolous lawsuits, the clarification and restriction of statutes of limitations on 10(b)(5) actions, and the establishment of proportionate liability. Should legislative modification of *Central Bank* be seriously considered, the Dodd-Domenici bill contained what could be an acceptable version of aiding and abetting. According to an article appearing in the *National Law Journal* (1994) a nonpublic portion of the Dodd-Domenici bill provided for aiding and abetting liability inn instances involving the conscious disregard of potential 10(b) violations by professional advising a securities issuer (Donovan 1994).

CONCLUSION: CENTRAL'S LASTING IMPRESSIONS

The significance of *Central Bank* extends well beyond its application to aiding and abetting. As with the 1993 Supreme Court decision in *Reves, Central Bank* is based on strict statutory construction. Although, the Court expressly indicated that the decision was not decided on policy grounds, language in the majority opinion suggests that the Court favors policy arguments averse to expansive standards of liability. By rejecting the implication of rights, the decision limits the potential for judicial expansion of theories of liability not expressly authorized by statute. Strict statutory construction helps clarify the interpretation of statutes, thus enhancing the ability of courts to apply the law in a fair and consistent manner. While adoption of strict statutory construction may result in Congress enacting legislation allowing for aiding and abetting and more expansive theories of liability, in light of the November 1994 elections there appears little likelihood that Congress will take such action absent it being part of a compromise bill.

Looking solely at the express language of a statute in order to determine its construction raises uncertainty over the vitality of many long-standing aspects of securities law. For example, should recklessness now be rejected as sufficient to support 10(b)(5) liability? The language of *Central Bank* suggests that it should.

Central Bank has already had an important impact on litigation involving accountants. The decision narrows the options under which actions against accountants may be brought. As a consequence certain actions that would have been raised against accountants as aiding and abetting claims will likely be forgone. In many instances it is expected that alternative theories will be asserted by both private litigants and the SEC to replace aiding and abetting. The success of these alternative theories remains to be seen.

At present legislative overturning of *Central Bank* does not appear imminent. By relying on strict statutory construction *Central Bank* has made it evident that it is the legislature and not the judiciary that must establish legislation. Although expressly not basing its decision on policy factors, the Court set forth policy arguments which should be taken into consideration by Congress in the future should it confront the task of establishing parameters for accountant liability.

REFERENCES

Abell v. Potomac Insurance Co. 858 F.2d 1104 (5th Cir. 1988); vacated in part on other grounds, sub nom. *Fryar v. Abell*, 109 S. Ct. 322 (1989); *Abell v. Wright, Lindsey and Jennings.* 109 S.Ct. 3242 (1989).

American Continental Corporation/Lincoln Savings and Loan. 794 F.Supp. 1424 (D. Ariz. 1992).

Brennan v. Midwestern Union Life Ins. Co. 286 F.2d 702 (N.D. Ind. 1966); aff'd 417 F.2d 147 (7th Cir. 1969); cert. denied 397 U.S. 989 (1970).

Bureau of National Affairs. 1994a. Securities cases repled in view of Supreme Court decision, McLucas says. *Daily Report for Executives*, November 8.

————. 1994b. Securities, Levitt says no plans to seek legislation to reform litigation. *Daily Report for Executives*, November 7.

Central Bank of Denver, N.A. v. First Interstate Bank of Denver, N.A. 114 S.Ct. 1439 (1994), *First Interstate Bank of Denver, N.A. v. Pring*, 969 F.2d 891 (10th Cir. 1992), *First Interstate Bank of Denver, N.A. v. Kirchner Moore and Company*, D.C. No.s 89-F-1250, 89-F-1006, slip op. (D. Colo. 1990).

Cleary v. Perfectune, Inc. 700 F.2d 774 (1st Cir. 1983).

Commodity Exchange Act Section 25(a).

Cortec Industries, Inc. et al. v. Sum Holding, L.P., et al. 858 F. Supp. 34 (S.D. N.Y. 1994).

Donovan, K. 1994. Bill would reverse Central Bank. *National Law Journal*, August 29, p. 29.

18 U.S.C.A. 1961-1968.

Ernst and Ernst v. Hochfelder. 425 U.S. 185 (1976), rev'g 503 F.2d 1104 (7th Cir. 1974).

Farlow v. Peat, Marwick, Mitchell and Co. 956 F.2d 982 (10th Cir. 1992).

Fine v. American Solar King Corp. 919 F.2d 290 (5th Cir. 1990).

Fischel, D.R. 1981. Secondary liability under section 10(b) of the Securities Act of 1934. *California Law Review* 69:80.

Gilmore v. Berg. 820 F. Supp. 179 (D.N.J. 1993).

H.R. 417 (proposed by Rep. Tauzin (D-La.).

Hayes, et al. v. Arthur Young and Company, et al. 34 F.3d 1072 (9th Cir. 1994).

Hearings concerning *Central Bank of Denver* decision before Subcommittee on Securities of the Senate Committee on Banking, Housing and Urban Affairs, 103rd Cong., 2d Sess. FDCH Cong. Testimony at 4 (May 12, 1994).

Herman and MacLean v. Huddleston. 459 U.S. 375, 379 n.5 (1983).

IIT v. Cornfeld. 619 F.2d 909 (2nd Cir. 1980).

In Re: Kendall Square Research Corporation Securities Litigation. Action No. 93-12352-EFH (N.D. Mass. 1994).

In Re: Medeva Securities Litigation. Fed. Sec. L.Rep. CCH P98,323 (1994).

I.R.C. Section 6701.

Kadish, S. 1985. Complicity cause and blame: A study in the interpretation of doctrine. *California Law Review* 73:332.

K and S Partnership v. Continental Bank, N.A. 952 F.2d 971 (8th Cir. 1991), cert. denied 112 S.Ct. 2993 (1992).

Levine v. Diamanthuset, Inc. 950 F.2d 1478 (9th Cir. 1991).

Monsen v. Consolidated Dressed Beef Co. 579 F.2d 793 (3rd Cir.), cert. denied 439 U.S. 930 (1978).

Moore v. Fenex, Inc. 809 F.2d 297 (6th Cir.), 483 U.S. 1006 (1987).

Morin, 832 F.Supp. 97 (1993).

Nolte v. Pearson. 994 F.2d 1311 (8th Cir. 1993).

Reves v. Ernst and Young. 61 USLW 4207 (1993).

Roberts v. Peat, Marwick, Mitchell and Co. 857 F.2d 646 (9th Cir. 1988), cert. denied 110 S.Ct. 56 (1989).

Rolf v. Blyth, Eastman, Dillon and Co. 570 F.2d 38 (2nd Cir. 1978), cert. denied 439 U.S. 946 (1978).

Ruder, D.S. 1994. The future of aiding and abetting and Rule 10b-5 after Central Bank of Denver. *The Business Lawyer* 49: 1479, 1481-1482.

S. 1976 (proposal of Senator Dodd (D-Conn.) and Senator Domenici (R-N.M.).

S. 2306 (proposal of Senator Metzenbaum D-Ohio).

Sassoon v. Altgelt 777 Inc. 822 F. Supp. 1303 (N.D. Ill. 1993).

Schatz v. Rosenberg. 943 F.2d 485 (4th Cir. 1991), cert. denied 112 S.Ct. 1475 (1992).

Schlifle v. Seafirst Corp. 866 F.2d 935 (7th Cir. 1989).

Schneberger v. Wheeler. 859 F.2d 1477 (11th Cir. 1988), cert. denied 490 U.S. 1091.

SEC discussing possible legislation in wake of aiding and abetting decision. 62 *Banking L. Rep.* (BNA) (April 29, 1994, May 9, 1994 and May 16, 1994).

Securities Act Release No. 5008 (Fed. Sec. L. Rep. CCH P77,913 (September 24, 1970).

SEC v. Militano. 1994 U.S. Dist. LEXIS 14485 (S.D.N.Y. 1994).

SEC v. Seaboard Corp. 677 F.2d 1301 (9th Cir. 1982).

Seligman, J. 1994. The implications of Central Bank. *The Business Lawyer* 49:1429, 1433.

Teague, et. al. v. Bakker, et al. 35 F.3d 978 (4th Cir. (1994)).

Vosgerichian v. Commodore Int'l, et al. 862 F. Supp. 1371 (E.D. Pa. (1994).

Wager, L.K., and J.F. Failla. 1994. *Central Bank of Denver, N.A. v. First Interstate Bank of Denver, N.A.*—the beginning of an end, or will less lead to more. *The Business Lawyer* 49: 1451, 1452-1453.

REFORMING ACCOUNTANTS' LIABILITY TO THIRD PARTIES AND THE PUBLIC INTEREST

Zhemin Wang

ABSTRACT

Accountants' liability to third parties for ordinary negligence has expanded dramatically in the last two decades. While accountants argue that the legal system has excessively burdened the accounting profession, proponents of the current liability system insist that such massive liability is necessary to protect the public interest through increased vigilance, and that any legal reform would only serve the self-interest of accountants and will leave the public with little protection. The analysis of this study indicates that the current liability system regarding accountants' liability to third parties for ordinary negligence unfairly overpenalizes accountants and is economically inefficient. Furthermore, contrary to the claim that it protects the public interest, the analysis suggests that third-party users and the general public whom the system is supposed to protect are the

Research in Accounting Regulation, Volume 9, pages 163-179.

consequential victims of the unfair legal attack on accountants. It is concluded that reforming the liability system to allow a more equitable risk sharing would not only make the liability system fairer and more economically efficient, but also serve the best interest of third-party users and the general public.

INTRODUCTION

As a growing number of jurisdictions adopted some variation of the more liberal interpretation of the privity requirement in cases against accountants by third-party users, accountants' liability to third parties for ordinary negligence has expanded dramatically in the last two decades. Consequently, in spite of the lack of evidence that accountants are any less careful or capable than they were ten years ago, litigation against accountants by third parties increased in logarithmical proportions since the 1980s (AICPA 1992; O'Malley 1993; Lochner 1992; Parker and Baliga 1988; Pae 1990). As accountants are hammered with lawsuits, judgment, and settlement, the profession itself is at risk. It has become apparent that litigation has become the most far-reaching and pervasive problem that the profession faces (Statement of Position by the Six Largest Public Accounting Firms 1992; Allen 1990; Goldwasser 1988; Hill et al. 1993; Miller 1988; Woolf 1985).

Different schools of thought currently exist on whether and to what extent an accountant's duty of care in the preparation of an independent audit of a client's financial statements extends to third-party users. Proponents of the current legal system contend that imposing massive liabilities on accountants for ordinary negligence is necessary to protect the public interest through "increased vigilance" (Doucet 1993). Furthermore, it is argued that limiting accountants' liability to third parties for ordinary negligence only serves accountants' self-interest, and will "leave the public with little protection." On the other hand, those against the recent expansion of accountants' liability to third parties argue that allowing a virtually unbounded class of claimants to sue accountants for ordinary negligence has excessively burdened the accounting profession, and call for legal reforms to "restore equity and sanity to the liability system" and to provide reasonable assurance that the public accounting profession will be able to continue to meet its public obligations (e.g., see AICPA 1992).

This study attempts to address the relationship between reforming accountants' legal liability to third parties for ordinary negligence and the public interest by examining the fairness and efficiency of the current legal system in the auditing context, as well as its impact on public interest. First, the fairness of the recent expansion of accountants' liability to third parties for ordinary negligence is examined. The analysis indicates that the current legal system unfairly overpenalizes slightly negligent accountants, and underpenalizes primary perpetrators. Furthermore, the threat of prohibitive legal costs, damage to reputation, the unpredictable nature of the outcome of a jury trial, and the prospect of having to pay all damages as a consequence of joint and several liability compels honest and competent accountants into settlements with meritless claims. Second, the economic efficiency of the current legal system in auditing context is examined. An efficient legal system in the auditing context is defined as one that maximizes the wealth of the society both by effectively deterring the occurrence of financial losses due to fraudulent financial reporting and by minimizing the cost of preventing and settling such losses (Epstein and Spalding 1993). The analysis suggests that the current legal system which concentrates liability exposure on a single minor defendant—the accountant—is ineffective in deterring the occurrence of fraudulent financial reporting and tends to increase both the frequency and cost of litigation. Finally, the analysis of the impact on third-party users and the general public of imposing an unfair burden on accountants indicates that third-party users and the general public, whom the system is supposed to protect, are consequential victims of the legal attack on accountants. In summary, it is concluded that limiting accountants' liability to third parties to allow a more equitable risk sharing would not only make the system fairer and more economically efficient, but also serve the best interest of third-party users and the general public.

The remainder of this study is organized as follows: the next two sections examine the fairness and economic efficiency of the current legal system in auditing context. The fourth section analyzes the impact on public interest of the legal attack on the accounting profession and the fifth section summarizes the conclusions.

FAIRNESS OF ACCOUNTANTS' LIABILITY
TO THIRD PARTIES

Accountants may be held liable to third-party users of audited financial statements for ordinary negligence under either the statutory law (primarily Section 10(b) of the Securities Exchange Act of 1934 and Rule 10(b)-5 of the Securities Exchange Commission) or the common law. In *Ernst and Ernst v. Hochfelder*,[1] the Supreme Court held that Rule 10(b) of the Securities Exchange Act of 1934 was inapplicable for holding accountants liable for mere negligence. Therefore, in the absence of a provable intent to deceive, manipulate, or defraud, no liability was imposed on alleged injurious reliance on negligently audited financial statements.[2] Individuals who allegedly suffered financial losses from reliance on negligently audited financial statements had to look to the common law for recovery.

In early common law, the public accountants' legal liability to third parties for ordinary negligence was based primarily on the concept of privity. Privity of contract means that the rights or obligations that exist under a contract are between the original parties to that contract, and failure to perform with due care results in a breach of that duty to only those parties. There are, however, inconsistent approaches across jurisdictions regarding the privity requirement, and courts through the years have tended to be less insistent upon privity in cases against public accountants (Jaenicke 1977; Epstein and Spalding 1993). There are, in general, three different approaches currently adopted by various jurisdictions regarding the privity requirement.

1. *Traditional View: The Privity Rule.* An early precedent of auditors' liability to third parties was established in the famous *Ultramares Corp. v. Touche*,[3] which ruled that auditors could not be held liable to nonclients for negligence because there was no privity. For a third party to hold auditors liable, he/she has to show fraud or gross negligence. The doctrine of privity had, thus, effectively served to shield accountants from third-party liability. Today, there are about twelve states still following the privity requirement of the *Ultramares* case.

2. *The Restatement (Second) Approach.* A federal district court decision[4] and the Restatement (Second) of the Law of Torts[5] are instrumental to the development of this alternative approach to

the traditional privity view. This approach represents a broad departure from the traditional requirement of privity in that it expanded auditors' liability to users who were known to (or should have been foreseen by) the accountant, or members of a class of users known to the accountants. The Restatement (Second) approach has quickly replaced the *Ultramares* approach in many states. Currently, approximately 18 states are following a variation of the Restatement (Second) approach.

3. *Foreseeable Users Approach.* In a New Jersey case, *H. Rosenblum, Inc. v. Adler,*[6] the court further departed from the *Ultramares* approach and expanded accountants' liability to all reasonably foreseeable third parties. Under this approach, a third party only needs to prove that the accountant should have expected the nonclient to make use of the audited financial statements; therefore, the class of the allowable claimants is virtually unbounded. The emergence of the foreseeable users approach has dramatically increased the exposure of auditors to massive liability for negligence. Although this approach currently is still a minority, it has inflicted significant damage to the accounting profession: auditors have been held liable for compensatory damages equal to the total losses suffered by creditors, or even guarantors of the creditors of an insolvent company (Hill and Metzger 1992; Tucker and Eisenhofer 1990; Lochner 1993). In addition, such liability extended to capital losses incurred by equity investors. Because auditors' liability to third parties sounded in tort, some juries awarded punitive damages to third-party plaintiffs for amounts that were wildly out of proportion to injury caused (see Table 1).

Because cases against accountants typically involve insolvent businesses, the primary perpetrators are usually judgment proof (Palmrose 1987). Consequently, under the joint and several liability system, the slightly negligent accountant (say, 1% at fault under a comparative fault determination) could be assigned to bear 100% of third-party users' damages. Proponents of the current liability system argue that in choosing which party on whom to assign the risk of an insolvent primary perpetrator, it is better to punish a proven wrongdoer than the innocent victim.[7] This argument clearly has its merit in cases involving "physical force" in which the victim is powerless in preventing the injury; the causal relationship is evident, and the urgency to compensate victims of personal injury is high. Cases against accountants by third parties clearly lack these characteristics.

Table 1.   The Legal Environment of Accountants' Liability
To Third Parties for Ordinary Negligence[1]

Privity Jurisdictions	Restatement Second Jurisdictions	Foreseeable Users Jurisdictions
Alabama, Idaho, Nebraska, Pennsylvania, New York, Delaware, Colorado, Indiana, Arkansas, Illinois, Kansas, and Utah	Florida, Georgia, Iowa, Kentucky, Louisiana, Michigan, Minnesota, Missouri, Montana, New Hampshire, North Carolina, North Dakota, Ohio, Rhode Island, Tennessee, Texas, Washington, and West Virginia	New Jersey, Wisconsin, Mississippi, and California[2]

Notes: [1] Currently, there are three approaches across jurisdictions to the question whether accountants should be subjected to liability to third-party users for ordinary negligence, namely the privity jurisdictions (about 12 states), Restatement Second jurisdictions (about 18 states), and foreseeable users jurisdictions (about 4 states). Although these three schools of thought are commonly recognized, there are some variations within each school. In addition, some states are still undecided on this issue.
[2] In a recent California case, the Supreme Court of California rejected the foreseeable users approach in favor of the Restatement Second Approach [see *Bily v. Arthur Young & Co.*, 834 P.2d 745 (Cal. 1992)].

First, cases against accountants involve losses that are purely pecuniary in nature. The urgency of the policies to compensate victims for financial losses is clearly not as intense as for victims of personal injury. Second, the relationship between auditor's conduct and third-party's financial losses is generally attenuated, and the third-party's reliance on audit report may be easily fabricated. Investment and credit decisions are by their very nature complex and multifaceted. Although an audit report might play a role in such decisions, reasonable and prudent investors and creditors will dig far deeper in their "due diligence" investigations than the surface level of an auditor's report. The ultimate decision to lend or invest is often based on numerous business factors such as the products or services of the company, its market environment, and its management personnel, all of which the auditor has no expertise in or control over. Yet, when the business failed, the plaintiff's attorney would try to convince the jury that the plaintiff relied solely on the auditor's report. Clearly, the relationship between plaintiff's financial losses and the audit report is something less than a "close connection." Furthermore, given that the auditor has no control over the distribution of the audit report once it reaches the hands of the client, it is difficult for the auditor to prove that any particular plaintiff did not receive, read, or rely on the report. Whether a plaintiff first encountered the report before making the investment or credit decision (when reliance could reasonably be inferred) or in the office of a lawyer after the company failed (when no such reliance could be drawn) is not readily susceptible to verification from any unbiased source. Finally, third-party victims who allegedly suffered financial losses from relying on negligently audited financial statements are generally aware that their loans or investments are subject to numerous risks including those flowing from a world of imperfect information. These third-party users, unlike the victims of "physical force," are not powerless in controlling their investment or credit risks. On the contrary, they have many options to hedge against or control these risks (e.g., diversification or using their contracting power to privately order the risk), and knowingly take those risks for a return (a potentially unlimited return for equity investors). Accountants, on the other hand, have no additional interest in the company they audit beyond a fee for their service.

In summary, in contrast to the "presumptively powerless victim" of "physical force," third-party users in an audit negligence case have many other options to prevent their financial losses; the relation between third-party's financial losses and the auditors' conduct is attenuated; and the urgency of the policies to compensate victims for financial losses is not as intense as for victims of personal injury. Given the above discussion, it does not seem inherently fair to hold accountants liable to a virtually unlimited number of claims of doubtful merit, to punish slightly negligent accountants beyond what they are responsible for, and to assign accountants to bear the entire risk of an insolvent primary perpetrator.[8] It is rather ironic that under the foreseeable users approach accountants bear 100% of the risk of insolvent primary perpetrators of failed businesses while the capital venturers bear no risk, at least until the accountants become personally bankrupt. Clearly, some sort of reform that would limit accountants' liability to third parties and would allow an equitable risk sharing is called for.

Some justify the burden imposed on a slightly negligent accountant by saying that it is better to punish a proven wrongdoer than an innocent victim. This maxim may be applicable to personal injury cases, where the juries are generally familiar with the instrumentalities that cause the injuries and recognize many of the circumstances. Consequently, common sense judgments are possible, or at least the verdicts seem to be accepted by the public (Epstein and Spalding 1993). In auditing cases, however, neither juries nor judges are likely to be as familiar with the generally accepted accounting and auditing principles applied by the auditors, with the inherent limitations in applying them, nor with the alternatives that might have been applied and their limitations. This unfamiliarity of juries and judges with the technical background in complex litigation against accountants creates a distinct possibility for honest and competent accountants to be found negligent by the jury and punished by the courts. After all, when a firm fails, it is hard to defuse the plaintiff's assertion that "accountants must be negligent, else how could these nice people lose their money?" (Lochner 1993).

In a recent California case,[9] the third-party plaintiff had not received, or read the audit report and, therefore, could not possibly have relied on it in making his highly speculative investing decisions. The jury nonetheless returned a verdict in his favor, allowing him to recover from the auditor his financial loss when his speculation ultimately was not materialized. Although the verdict was later reversed by the Court of Appeal, the case nevertheless demonstrates the difficulties for accountants in defending themselves in cases by third parties and the uncertainties of jury trial. This uncertainty, together with the lack of an upper limit to accountants' liability, have forced many honest and competent accountants into settlements on meritless claims (Lochner 1993; Statement of Position 1992). Furthermore, because plaintiffs theoretically may recover all of their losses from an auditor, proven even slightly negligent, the high stakes in these complex litigations make the trial expensive and create tremendous financial burdens on innocent accountants even if they eventually prevail in the trial. As pointed out by Robert Levine, the former CEO of the bankrupted Laventhol & Horwath (the nation's seventh largest accounting firm), "it wasn't the litigation we would lose that was the problem. It was the cost of winning that caused the greatest part of our financial distress." Clearly, the prohibitive legal costs and the lengthy and disruptive discovery process had influenced honest and competent accountants to enter into settlements on meritless claims, which, in turn, has created yet another incentive for third parties to file even more meritless cases against accountants. This snowball effect explains the geometric increase in cases against accountants by third parties in the last decade, despite the lack of evidence that accountants are any more culpable than they were ten years ago.

In summary, the analysis suggests that imposing on accountants an unlimited liability to an unbounded class of allowable claimants for ordinary negligence not only unfairly overpenalizes slightly negligent accountants by assigning them to bear all the risk of an insolvent primary perpetrator, but also indiscriminately punishes honest and competent accountants by forcing them into settlements with plaintiffs on meritless claims.

ECONOMIC EFFICIENCY OF ACCOUNTANTS' LIABILITY TO THIRD PARTIES

An economically efficient legal system in auditing context should maximize the wealth of the society by minimizing the frequency of loss occurrence and the cost of loss prevention. Proponents of the current liability system argue that imposing massive liabilities on accountants effectively deters the occurrence of fraudulent financial reporting and is, therefore, economically efficient.[10] This section examines whether the current legal system which imposes massive liabilities to accountants indeed effectively deters the occurrence of third-party's financial losses from fraudulent financial reporting and minimizes the cost of litigation.

As a matter of commercial reality, audits are performed in a client controlled environment. The client typically prepares its financial statements; it has direct control and assumes primary responsibility for their contents. Because of the inherent time limitation, the client necessarily furnishes the information base for the audit. Clearly, the primary responsibility for preventing fraudulent financial reporting rests with the auditor's clients, and the auditor's role in the financial reporting process is only secondary. It is evident that auditor's clients are either the primary perpetrator or the less costly one to prevent losses from fraudulent financial reporting. Consequently, in order to effectively deter the occurrence of losses from fraudulent financial reporting, the optimum liability rule should be to assign liability only to auditors' clients. This is consistent with the philosophy underlying the traditional privity rule established by the *Ultramares* case.[11]

Cases against accountants typically involve insolvent primary perpetrators. Consequently, the elimination of the privity requirement in cases against accountants by third parties implies that both equity and debt investors can sue accountants for ordinary negligence and can recover, as a result of the joint and several liability, the entire loss from a slightly negligent accountant. Because plaintiffs can recover all of their financial losses from the slightly negligent accountant, plaintiffs and their attorneys have little incentive to go after the judgment proof primary perpetrators (Birnbaum 1986). As a common practice, plaintiffs' attorneys usually settle with the primary perpetrators at a fraction of what these parties should pay. The attorneys then pursue the case against the "deep pockets" accountants, who as a result of joint and several liability are exposed for 100% of the damage even if found to be only 1% at fault. It is evident that while accountants are overpenalized and overdeterred, the primary perpetrators are underpenalized and underdeterred. Clearly, the deterrence objective of the legal system is unlikely to be achieved by imposing massive liabilities on accountants.

Furthermore, as a sound economic and social policy, third parties should be encouraged to rely on their own prudence, diligence, and contracting power, as well as other informational tools, in preventing investment losses. This kind of self-reliance promotes sound investment and credit practices and discourages the careless use of monetary resources. Under the current liability system, however, third parties are allowed to recover financial losses from the auditor who becomes, in effect, an insurer of not only financial statements, but of bad loans and investment in general (Hill and Metzger 1992; Allen 1990). It is obvious that allowing third-party users to transfer their losses to accountants provides these third parties little incentive to take necessary precautions to prevent such losses; thus, rather than minimizing the frequency of loss occurrence, the current legal system exacerbates it.

Finally, the elimination of the privity requirement in cases against accountants by third parties for ordinary negligence has significantly raised the stakes in such litigation. As the stakes to auditors and the third-party users are raised, both parties are more likely to invest more in the contest itself. Auditors faced with 100% of third-party users' damage from an insolvent client are more likely to fight hard and expensively to avoid liability altogether than they would if faced only with their equitable portion of the damages. Similarly, plaintiffs and their attorneys who see a possible massive recovery are also more likely to incur a larger cost in litigation than they would if a small prize were promised. Clearly, the current legal system regarding accountants' liability to third parties, which concentrates liability exposure to a single party—the accountants—because of their "deep pockets," increases both the likelihood and the cost of litigation against accountants.

In summary, the current legal system regarding accountants' liability to third parties overpenalizes and overdeters accountants who play only a secondary role in the financial reporting process and underpenalizes and underdeters primary perpetrators. Furthermore, the system does not promote third-party's self-reliance in preventing investment and credit losses, and it tends to increase both the likelihood and the cost of litigations against accountants. Such a system is not only ineffective in deterring the occurrence of third-party users' financial losses, but also likely to increase the total cost of litigation.

PUBLIC INTEREST AND ACCOUNTANTS' LIABILITY TO THIRD PARTIES

This section examines the direct impact of expanding accountants' liability on third-party users and the general public. Proponents of the current liability system argue that the massive liability imposed on accountants enhances the quality of audit service and, therefore, can prevent future third-party financial losses. For example, the New Jersey Supreme Court reasoned: "The imposition of a duty to foreseeable users may cause accounting firms to engage in more thorough reviews. This might entail setting up stricter standards and applying closer supervision, which would tend to reduce the number of instances in which liability would ensue."[12]

However, in view of the inherent dependence of the auditor on the client in performing the audit and the labor-intensive nature of auditing, it is highly unlikely that audits can be done in ways that would yield significantly greater accuracy without significant increases in cost and other disadvantages. Moreover, an audit report is not a simple statement of verifiable fact that, like the weight of the load of bean in the *Glanzer v. Shepherd* case,[13] can be easily checked against uniform standards of indisputable accuracy. Rather, an audit report is a professional opinion based on numerous and complex factors which are subject to the auditor's interpretation and application of hundreds of professional standards, many of which are broadly phrased and readily subject to different constructions. Thus the audit report is the final product of a complex process involving discretion and judgment on the part of the auditor at every stage. Using different initial assumptions and approaches, and different sampling techniques, few audits would be immune from criticism. Obviously, if the auditor will inevitably be singled out and sued when the client goes into bankruptcy regardless of the care or detail of the audits, the auditor would rationally respond to increased liability by simply reducing audit service in industries where the business failure rate is high ("Lawsuit fears..." 1990).[14] As pointed out by a legal economist: "The deterrent effect of liability rules is the difference between the probability of incurring liability when performance meets the required standards and the probability of incurring liability when the performance is below the required standards. Thus, the stronger the probability that liability will be incurred when performance is adequate, the weaker is the deterrent effect of liability rules. Why offer a higher quality product if you will be sued regardless whenever there is a precipitous decline in stock price?"[15]

Those in favor of assigning massive liabilities to accountants for ordinary negligence generally assume that accountants have little difficulty passing this liability burden to their clients and therefore predict an efficient risk spreading. For example, in *Rusch Factors, Inc v. Levin*,[16] the court states,

> Why should an innocent party be forced to carry the weighty burden of an accountants' professional malpractice? Isn't the risk of loss more easily distributed and fairly spread by imposing it on the accounting profession, which can pass the cost of insuring against this risk on to its customers, who can in turn pass the costs on to the entire consumer public?

Unfortunately, this assumption is dangerously erroneous. In reality, the heavy competition for clients and the existence of "go bare" auditing firms make it difficult to raise auditing fees (Francis and Simon 1987; Palmrose 1986; Ettredge and Greenberg 1990; Schatzberg 1990). Moreover, insurance coverage for auditors is either prohibitively costly or not available altogether (AICPA

1992). The risk sharing mechanism based on this false assumption is, therefore, unlikely to be optimal.

Because of the indeterminate nature of auditors' liability to third parties, insurance companies have no scientific basis for calculating the litigation risk against accountants, and have generally responded to the major cases against accountants by either increasing the rates by thousands of percents in a short period of time or dropping the entire line of auditing insurance altogether (AICPA 1992). Audit firms whose liability insurance is either prohibitively costly or not available altogether may be forced to "go bare" (Epstein and Spalding 1993). The "go bare" auditors are either uninsured or greatly underinsured, and make themselves judgment proof by having little assets under their own names. Those in favor of the current liability system argued that expanding accountants' liability to third parties would expose accountants to greater risk for ordinary negligence, which, in turn, would provide incentives for accountants to incur additional cost on quality control commensurate with the increased risk. Because "go bare" auditing firms are judgment proof, however, expanding the allowable claimants against them will not expose these firms to greater risk and, thus, will not necessarily make them incur additional quality control cost.

When sued by third parties, these judgment proof auditing firms will shift the risk of insolvent clients right back to the third parties. Clearly, forcing accounting firms to "go bare" will not provide third parties the extra protection claimed by those in favor of expanding accountant's liability to third parties. In summary, the above discussion casts doubt on the predicted positive relationship between accountants' legal exposure and audit quality. It is worth noting, however, that "going bare" does not mean that these firms are attempting to avoid their obligations to the public or to reduce their commitment to quality audit service. Rather, it is simply a rational, and many times the only available, option to accountants in responding to the excessive legal burden imposed by the current liability system.

The dramatic expansion of accountants' liability to third parties exposes auditors choosing not to "go bare" to significantly greater risk and will undoubtedly make these firms spend more on quality control. However, the competitiveness of the auditing market and the existence of "go bare" firms make it difficult for firms that are reluctant to "go bare" to raise auditing fees to compensate their additional cost on quality control and their increased insurance premiums (Rubin 1988; Simon and Francis 1988; Schatzberg 1990; Somer 1990). In order to survive, they will have to either drop their risky clients altogether or become more conservative in their examination of clients and in rendering opinions, a response similar to the practice of defensive medicine.[17] This means companies in more risky industries such as the high-tech industry, the growth companies, and the less established companies may be unable to obtain quality auditing services, which, in turn, would hurt the third-party users. In addition, more unnecessary tests will be performed and more qualified opinions will be rendered, thus making it more difficult for the high-tech firms, growth firms, and small firms to raise capital.[18] Because these firms provide the most employment and opportunity for economic growth, the national economy, the general public, and the society at large are the consequential victims of the unfair legal attack on accountants.

In summary, contrary to the claim that expanded liability increases audit quality, the analysis suggests that decreased availability and increased cost of quality audit service are the more likely results. Opponents of legal reform warned of the risk of having accountants underdeterred because, they claimed, accountants may have no incentive to perform their duty of protecting the public interest. The above discussion, however, indicates that overpenalizing and overdeterring accountants can be equally harmful to the public interest, and suggests that reforms that would allow a more equitable risk sharing are urgently needed.

Finally, it is worth noting that the current liability system not only makes third-party users more likely to suffer from decreased availability and increased cost of quality audit service, but also makes it less likely for the true victims to be appropriately awarded. In order to create an overwhelming pressure on innocent accountants to settle, plaintiffs' lawyers have a strong incentive to bring as many cases as possible without regard to their relative merits and to include as many defendants as possible without regard to their degree of fault. The lawyers then settle these cases at a fraction of the alleged damages and typically receive 30% of the settlement plus expenses. The true victims, on the other hand, receive on average 5% to 15% of their damages, and are awarded no more than so-called "professional" plaintiffs and speculators trying to recoup investment losses (O'Malley 1993; Lochner 1993). By allowing virtually unlimited claimants with doubtful merit to sue accountants, the current legal system, in effect, reduces the probability for true victims to be appropriately awarded.

CONCLUSIONS

This study examined the relationship between reforming accountants' liability to third parties and the public interest. Contrary to the claim that such a reform serves only accountants' self-interest at the expense of the public, the analysis of this study indicates that reforming accountants' liability to third parties would make the legal system fairer, more economically efficient, and better serve the public interest.

The analysis was focused on the deficiencies of the current legal system, and concluded that the excessive legal liability to third parties imposed on accountants under the current system is unfair, economically inefficient, and does not serve the best interest of the public. It is suggested that some reform of the current legal system is urgently needed. However, the most effective method of limiting accountants' liability to third parties to allow a more equitable risk sharing requires more elaborate research studies, and is the subject for future research.

ACKNOWLEDGMENT

The author is grateful for comments on earlier versions of this manuscript by two anonymous referees, Michael Garrison, Robert Tucker, John Eichenseher, Terry Knoepfle, Carolyn Scheibelhut, Jim Hansen, and Bahman Bahrami.

NOTES

1. *Ernst & Ernst v. Hochfelder*, 425 U.S. 185 (1976).
2. Accountants may incur liability to third parties without a showing of fraud or gross negligence under section 18 of the Securities Exchange Act of 1934, 15 United States code section 78r, or section 11 of the Securities Act of 1933, 15 United States Code section 77k.
3. *Ultramares Corp. v. Touche*, 255 N.Y. 170, 174 N.E. 441 (1931).
4. See *Rusch Factors, Inc. v. Levin*, 284 F. Supp. 85 (D.R.I. 1968).
5. Restatement (Second) of Torts, Section 552 (Tentative Draft No.12, 1966).

6.   *H. Rosenblum, Inc. v. Adler*, 461 A.2d 138 (N.J. 1983).

7.   See *Rusch Factors, Inc. v. Levin*, op. cit.

8.   See *Brown v. Keill*, 580 P.2d 867 (Kan. 1978) wherein the court stated: "There is nothing inherently fair about a defendant who is 10% at fault paying 100% of the loss, and there is no social policy that should compel defendants to pay more than their fair share of the loss."

9.   See *Bily v. Arthur Young & Co.*, 834 P.2d 745 (Cal. 1992). In this case against Arthur Young & Co., one plaintiff had not received or read the audit report and, therefore, could not have relied on it in making his highly speculative warrant investment. The jury nonetheless returned a verdict in his favor.

10.   See *H. Rosenblum, Inc. v. Adler*, op. cit.

11.   See *Ultramares Corp. v. Touche*, op. cit.

12.   See *H. Rosenblum, Inc. v. Adler*, op. cit.

13.   See *Glanzer v. Shepard*, 233 N.Y. 236, 135 N.E. 275 (1992).

14.   There are evidences that large public accounting firms are restricting both the industries they will audit and the services they provide to some clients in an effort to reduce their legal risk.

15.   See Fischel, *The Regulation of Accounting: Some Economic Issues 52 Brooklyn Law Review* 1051, 1055 (1987). Some legal economists have argued that auditors cannot bear the virtually unlimited liability to third parties for ordinary negligence and that any attempt to force them to do so will result in lower audit quality as "deep pocket" auditors avoid risky industries; see also Hill and Metzger (1992).

16.   *Rusch Factors, Inc. v. Levin*, op. cit.

17.   Some legal economists have argued that even if auditors are able to fully price their legal risk and continue to audit risky clients, society still may not be better off because the cost of additional auditing and insurance will likely be far in excess of what the public is willing to pay (see, e.g., Goldwasser 1988).

18.   It has also been suggested that fear of legal liability has deterred auditors' innovation in developing ways to improve financial reporting (see Dopuch 1988; Sack 1988).

REFERENCES

Allen, J.M. 1990. Damage apportionment in accounting malpractice actions: The role of comparative fault. *Brigham Young University Law Review* 3: 949-985.

American Institute of Certified Public Accountants. 1992. *Survey of Accounting Firms*. New York: AICPA.

Birnbaum, S. 1986. Tort reform proposals analyzed. *The National Law Journal* (June 23): 18.

Dopuch, N. 1988. Implications of torts rules of the accountants' liability for the accounting model. *Journal of Accounting, Auditing and Finance* (Summer): 245-250.

Doucet, M. 1993. Tort reform: In whose interest? *Ethics in Accounting* (March): 4.

Epstein, M.J., and A.D. Spalding, Jr. 1993. *The Accountants' Guide to Legal Liability and Ethics*. Homewood, IL: Irwin.

Ettredge, M., and R. Greenberg. 1990. Determinants of fee cutting on initial audit engagements. *Journal of Accounting Research* (Spring): 198-210.

Fischel, D.R. 1987. The regulation of accounting: Some economic issues. *Brooklyn Law Review* 52: 1051-1056.

Francis, J.R., and D.T. Simon. 1987. A test of audit pricing in the small-client segment of the U.S. audit market. *The Accounting Review* (January): 145-157.

Goldwasser, D.L. 1988. Policy considerations in accountants' liability to third parties for negligence. *Journal of Accounting, Auditing and Finance* (Summer): 217-232.

Hill, J.W., and M.B. Metzger. 1992. Auditor liability and the S&L crisis: Shaping the future of the profession? *Annual Review of Banking Law* 11: 263-333.

Hill, J.W., M.B. Metzger, and J.W. Schatzberg. 1993. Auditing's emerging legal peril under the National Surety Doctrine: A program for research. *Accounting Horizons* (March):12-28.

Jaenicke, H. 1977. *The Effect of Litigation on Independent Auditors.* The Commission on Auditors' Responsibilities.

Lawsuit fears forcing auditors to cut services 1990. *The CPA Journal* (December): 6.

Lochner, P.R., Jr. 1992. Black days for accounting firms. *The Wall Street Journal,* May 22, p. A10.

————. 1993. Accountants' legal liability: A crisis that must be addressed. *Accounting Horizons* (June): 92-96.

Miller, S.H. 1988. Avoid lawsuits. *Journal of Accountancy* (September): 57-65.

O'Malley, S.F. 1993. Legal liability is having a chilling effect on the auditor's role. *Accounting Horizons* (June): 82-87.

Pae, P. 1990. Laventhol bankruptcy filing indicates liabilities may be as much as $2 billion. *The Wall Street Journal,* November 23, p. A4.

Palmrose, Z. 1986. Audit fees and auditor size: Further evidence. *Journal of Accounting Research* (Spring): 97-110.

————. 1987. Litigation and independent auditors: The role of business failures and management fraud. *Auditing* (Spring): 90-103.

Parker, R.M., and W.J. Baliga. 1988. Lessons from liability trends. *New Accountant* (January): 33-38.

Rubin, M.A. 1988. Municipal audit fee determinants. *The Accounting Review* (April): 219-236.

Sack, R.J. 1988. Discussion of implications of torts rules of the accountants' liability for the accounting model. *Journal of Accounting, Auditing and Finance* (Summer): 251-254.

Schatzberg, J.W. 1990. A laboratory market investigation of low balling in audit pricing. *The Accounting Review* (April): 337-362.

Simon, D.T., and J.R. Francis. 1988. The effect of auditor change on audit fees: Test of price cutting and price recovery. *The Accounting Review* (April): 255-269.

Somer, A.A., Jr. 1990. Is there a future for auditing. *The CPA Journal* (January): 7-8.

Statement of Position by the Six Largest Accounting Firms on the Liability Crisis Facing the Accounting Profession. *Journal of Accounting* (November): 19-23.

Tucker, M., and J. Eisenhofer. 1990. Accountants' liability: Negligent representation suits multiply in the wake of S&L crisis. *The National Law Journal* (June 25): 17-19.

Woolf, E. 1985. We must stem the tide of litigation. *The Accountant* (April): 18-19.

THE CHANGING PROFILE OF
THE AICPA:
DEMOGRAPHICS OF A MATURING PROFESSION

Stephen J. Young

ABSTRACT

In this paper I explore trends in the membership of the American Institute of Certified Public Accountants (AICPA) from 1887-1994. The paper has two primary goals to provide a single source of information on past membership data, and, on a preliminary basis, to explore trends in the membership. This exploration identifies three major trends in AICPA membership. First, after many decades of rapid increase, membership growth is slowing. This slowdown appears both in absolute growth rates and rates relative to key economic indicators. The second major trend is that members in corporate practice are now as large a membership category as those in public practice. The third major shift in the membership is gender-based. In the youngest age category, women outnumber men on the membership rolls. This ratio of women to men slowly declines as the age category increases, suggesting that women will play a much large role in the future leadership of the profession.

Research in Accounting Regulation, Volume 9, pages 181-198.

The paper is intended to be a preliminary investigation of available demographic data. Further work needs to be completed to provide more definite conclusions about AICPA demographic trends. Such work will enable us to better predict the future nature and direction of the profession of accountancy and plan accordingly.

INTRODUCTION

The profession of accountancy in the United States has changed dramatically since its formal inception in 1887. In that year, the American Association of Public Accountants consisted of 26 members (see Appendix). More than a century later, in 1994, the American Institute of Certified Public Accountants (AICPA) consisted of 318,829 members. This paper examines the growth of the profession over the intervening years. Further, it examines the current activities of the membership, revealing major changes in the composition of Institute members. Finally, the paper examines demographic and economic trends which have influenced the growth, comparing accountants to other professionals.

Early demographic work in any field is, of necessity, mainly archival in nature. Very little is currently known about the demographics of the accounting profession. The purpose of this paper is to establish a comprehensive point of reference for further, more detailed demographic analysis of the profession. The paper makes at least two contributions to the literature. First, the paper provides a data archive not accessible from any readily available source. Second, the paper examines trends in the data and draws four preliminary conclusions from the analysis.

First, the growth rate of AICPA membership has slowed dramatically over recent years. In fact, the membership is growing at its lowest rate since the Depression. Second, the structure of the membership has changed massively over the last several decades. Today, virtually equal proportions of the membership are in public and corporate practice. In 1970, public practice members outnumbered members in corporate practice by a ratio of roughly 2:1. Third, net growth in the public practice section since 1990 has been virtually nil. Membership growth rates are slowing relative to key demographic and economic indicators. The last observation is that women are becoming a larger fraction of the membership.

Table 1. Average Net Membership Increase

	5-year Averages			5-year Averages	
Period	Net New Members	Growth (%)	Period	Net New Members	Growth (%)
1887-89	2.0	9.60	1940-44	511.8	9.11
1890-94	6.8	18.03	1945-49	981.8	11.80
1895-99	7.6	10.50	1950-54	1,640.6	11.71
1900-04	11.2	13.23	1955-59	1,690.6	7.49
1905-09	57.2	67.71	1960-64	2,503.2	7.70
1910-14	15.8	4.46	1965-69	3,103.2	6.55
1915-19	38.8	3.34	1970-74	5,696.4	8.46
1920-24	329.8	20.18	1975-79	7,364.0	7.53
1925-29	131.0	9.98	1980-84	11,507.2	7.95
1930-34	(81.4)	−1.21	1985-89	11,005.0	5.53
1935-39	133.8	3.32	1990-94	4,639.2	2.17

All of these observations suggest a membership which is changing in fundamental ways, creating new challenges for the leadership of the Institute.

CURRENT AICPA MEMBERSHIP

Historical Growth of the Institute

The current membership of the AICPA has been reached by many years of sustained growth. Table 1 shows five-year average net

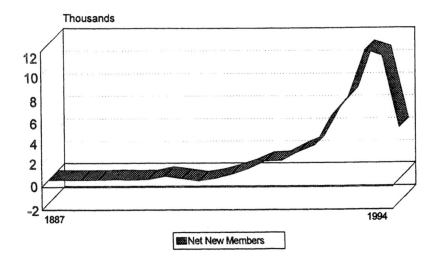

Figure 1. AICA Membership Growth 1887-1994

Table 2. AICPA Members Reported Activities, 1970-1994

Year	AICPA Membership	Public Practice (%)	Corporate Practice (%)	Education (%)	Government (%)	Retirees & Other[1] (%)
1970	75,381	61.6	31.3	3.3	3.8	N/A
1971	79,736	60.2	31.1	3.2	3.7	N/A
1972	87,562	60.5	32.9	3.1	3.5	N/A
1973	95,415	59.8	33.6	3.1	3.5	N/A
1974	103,863	60.0	33.6	3.0	3.4	N/A
1975	112,494	59.1	34.6	2.9	3.4	N/A
1976	121,947	58.5	30.9	2.9	3.4	4.3
1977	131,300	57.5	32.0	2.8	3.5	4.2
1978	140,158	57.6	31.9	2.8	3.4	4.3
1979	149,314	55.0	34.2	3.0	3.4	4.4
1980	161,319	54.1	35.5	2.9	3.3	4.2
1981	173,900	53.5	36.1	2.8	3.3	4.3
1982	188,706	52.5	37.6	2.5	3.2	4.2
1983	201,764	53.0	36.9	2.7	3.3	4.1
1984	218,855	51.5	38.4	2.7	3.3	4.1
1985	231,333	51.0	38.8	2.7	3.3	4.2
1986	240,947	49.1	39.5	2.8	3.2	5.4
1987	254,910	47.6	39.5	2.8	3.4	6.7
1988	272,479	46.5	39.6	2.7	3.6	7.6
1989	286,358	45.8	39.9	2.7	3.7	7.9
1990	295,633	44.5	40.4	2.7	3.7	8.7
1991	301,410	43.2	40.7	2.8	3.9	9.4
1992	308,280	42.6	40.6	2.4	4.1	10.3
1993	314,427	42.2	40.3	2.3	4.3	10.9
1994	318,829	41.3	40.9	2.4	4.4	11.0

Notes: [1] Percentages of the membership in this category are not available for the years 1970-1975. "Other" refers to members who report activities not readily identifiable with currently reported scope of service categories.

Source: AICPA Annual Reports, 1970-1994.

membership increases and growth rates. In every period except one, growth is positive. The only exception was in the period 1930-1934, at the height of the Great Depression. Since 1940, there has been a period of sustained membership growth.

In absolute numbers, this growth peaked in the 1980-84 period. By the 1990-94 period, net growth was the lowest since the 1965-69 period. This sharp drop in new membership can be more readily seen in Figure 1. Examination of growth rates tell a similar story. It is difficult to draw definitive conclusions from these observations. I now examine more detailed Institute data to more closely examine trends within the membership.

Sources and Occupations of AICPA Membership

The previously noted trend of declining growth rates is matched by a shift in the activities of the membership (refer to Table 2). In 1970, 61.6% of the membership was in public practice. By 1994, this number had decreased to 41.3% of the membership. Over this same period, members in corporate practice increased from 31.3% of the membership to 40.9% of the membership. Clearly, public practitioners are no longer a majority of the membership. In fact, there are now two substantial minorities of relatively equal size, public and corporate practitioners. Table 2 shows this shift over the years.

Proportions of members primarily involved in education or government have not changed substantially over the period. Finally, by 1994, 11% of the membership are retired or pursuing other[1] activities. In summary, the Institute has a membership composed of relatively equal minorities of public and corporate practitioners and a significant number of members pursuing varied activities.

Further, this trend does not seem likely to reverse itself in the future. The growth rate of the public practice section has long been below that of the corporate section (refer to Table 3). In the 1990s, growth in public practice membership been a total of 0.29% or a geometric average of 0.06% per annum. Effectively, these numbers represent a growth rate of zero. The growth rate of the corporate membership has also slowed in recent years to an average of 2.67%. The only consistently growing group of the membership is the "retirees and other" category.

Table 3. Growth Rates of AICA Members Reported Activities, 1970–1994

Year	AICPA Membership	Public Practice (%)	Corporate Practice (%)	Education (%)	Government (%)	Retirees and Other[1] (%)
1971	5.8	3.4	5.1	2.6	3.0	N/A
1972	9.8	10.4	16.2	6.4	3.9	N/A
1973	9.0	7.7	11.3	9.0	9.0	N/A
1974	8.9	9.2	8.9	5.3	5.7	N/A
1975	8.3	6.7	11.5	4.7	8.3	N/A
1976	8.4	7.3	-3.2	8.4	8.4	N/A
1977	7.7	5.8	11.5	4.0	10.8	5.2
1978	6.7	6.9	6.4	6.7	3.7	9.3
1979	6.5	1.7	14.2	14.1	6.5	9.0
1980	8.0	6.3	12.1	4.4	4.9	3.1
1981	7.8	6.6	9.6	4.1	7.8	10.4
1992	8.5	6.5	13.0	-3.1	5.2	6.0
1983	6.9	7.9	4.9	15.5	10.3	4.4
1984	8.5	5.4	12.9	8.5	8.5	8.5
1985	5.7	4.7	6.8	5.7	5.7	8.3
1986	4.2	0.3	6.0	8.0	1.0	33.9
1987	5.8	2.6	5.8	5.8	12.4	31.3
1988	6.9	4.4	7.2	3.1	13.2	21.3
1989	5.1	3.5	5.9	5.1	8.0	9.2
1990	3.2	0.3	4.5	3.2	3.2	13.7
1991	2.0	-1.0	2.7	5.7	7.5	10.2
1992	2.3	0.9	2.0	-12.3	7.5	12.1
1993	2.0	1.0	1.2	-2.3	7.0	7.9
1994	1.4	-0.8	2.9	5.8	3.8	2.3

Notes: [1] Percentages of the membership in this category are not available for the years 1970–1975. "Other" refers to members who report activities not readily identifiable with currently reported scope of service categories.

Source: AICPA Annual Reports, 1970–1994.

These trends in member activities suggest a future membership very different from what most people think the Institute is about. Given recent trends, members involved in corporate practice will soon outnumber the public practice membership.[2] This suggests a substantial shift in the Institute's priorities to accommodate its corporate practitioners. Corporate members are no longer a fringe group. They will soon be the largest single body of members. Combine this with a large group of members pursuing other activities, and the implications of such a diverse membership becomes of interest to policymakers. This diversity in member activities provides the Institute a challenge: satisfying the needs of *all* members in the next decade.

PROFESSIONAL MARKET TRENDS

Demographic Trends

Looking at broader demographic trends allows one to put the growth of the profession in perspective. This section generally uses the U.S. Census definition of practitioners in the various fields. The Census Bureau's definition of accountants is substantially broader than those accountants who are CPAs. For that reason, I report the AICPA membership for demographic comparison with the Census statistics on physicians, engineers, and lawyers. Of the four professions, accountancy is the only professional body which has a substantial numerical divergence of membership from the Census definition.

The purpose of this section is to illustrate two points: first, how the "the professions," especially accountants and engineers, have become a more important part of our modern information-based society, and second, the levelling of the prevalence of the professionals relative to the U.S. population and labor force. What follows is a sort of "saturation analysis." That is, the analysis attempts to detect trends in professional "density" over time and provide perspectives for predictions of future growth.

As we can see from Table 4, the ratio of members of the U.S. adult population for each professional has declined substantially since the turn of the century. These declines have been steady over the years (see Figure 2). CPAs are 817 times more prevalent in today's society than they were in 1900. On the other hand, engineers are 12.5 more prevalent,

Table 4. Number of U.S. Population[1] per Professional

	1900	1910	1920	1930	1940	1950	1960	1970	1980	1990
CPAs	526,120	61,280	51,858	17,595	17,791	6,771	3,215	1,865	1,121	644
Engineers	1,274	792	527	390	326	204	141	116	119	102
Lawyers	448	530	575	526	531	601	577	541	327	261
Physicians	369	401	484	540	576	552	561	549	400	328

Note: [1] U.S. population over the age of 16 years. Taken from U.S. Department of Commerce, Bureau of the Census of the Population, various years.
Comparisons of American Bar Association and American Medical Association membership suggest that these associations' membership statistics closely track those reported by the Bureau of the Census.

lawyers only 1.7 times more prevalent, and doctors 1.1 times more prevalent in the population than they were in 1900. Per capita membership in the medical and legal professions peaked in the 1940s and 1950s and has been declining to current levels ever since.

Table 5 statistics show a massive growth in the prevalence of accountants[3] relative to engineers, lawyers, and doctors and are consistent with the increased demand for information required in amodern economy.

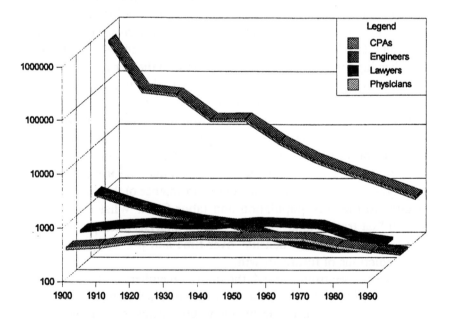

Figure 2. U.S. Population per Number of Professionals, Selected Years

Table 5. Number of U.S. Labor Force[1] per Professional

	1900	1910	1920	1930	1940	1950	1960	1970	1980	1990
CPAs	237,315	27,779	29,554	9,485	9,742	3,976	1,904	1,136	732	429
Engineers	575	359	301	218	178	120	83	71	76	68
Lawyers	202	240	327	294	291	353	342	329	209	174
Physicians	167	182	276	302	315	324	332	334	256	219

Note: [1] U.S. nonmilitary labor force. Taken from U.S. Department of Commerce, Bureau of the Census of the Population, various years.

Comparisons of American Bar Association and American Medical Association membership suggest that these associations' membership statistics closely track those reported by the Bureau of the Census.

A similar comparison of labor force data to the number of professionals in the economy leads to results comparable with the population statistics. The number of members of the labor force per accountant declined 553.9 times from 1900. The prevalence of engineers in the labor force increased 8.5 times and lawyers are only 1.2 times more common in the labor force than in 1900. The proportional number of doctors actually decreased to 76% of the 1900 per capita level. Referring to Figures 2 and 3, comparisons to population and labor force statistics tell similar stories. The graphs show the relative growth rate of all these professions levelling off, suggesting that some saturation has been achieved. This section reveals an important point: it appears that the relative number of accountants in the labor force is levelling off.

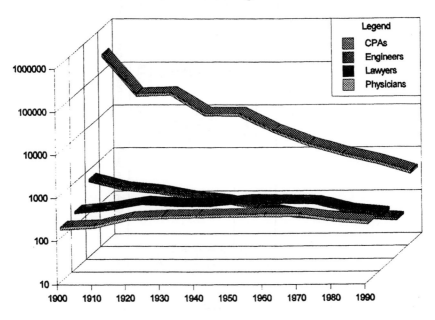

Figure 3. Members of the U.S. Labor Force per Professional Selected Years

Table 6. Decomposition of AIPCA Membership by Gender

Age Group	Male		Female		Sex Not Reported		Totals by Group	
	Number	Percent of Group	Number	Percent of Group	Number	Percent of Group	Number	Percent of Group
<26	2,049	48.88	2,089	49.84	54	1.28	4,192	1.30
26-35	55,246	56.96	34,127	35.18	7,628	7.86	97,001	30.20
36-45	82,392	70.33	24,240	20.69	10,525	8.98	117,157	36.48
46-55	51,127	79.61	7,291	11.35	5,809	9.04	64,227	19.99
56-65	18,863	86.31	1,294	5.92	1,699	7.77	21,856	6.81
>65	14,818	88.41	440	2.62	1,504	8.97	16,762	5.22
Totals	224,495	69.89	69,481	21.63	27,219	8.48	321,195	100.00

Source: AICPA Membership Demographics Report (unpublished, dated March 14, 1994).

Gender Trends

Also of interest is the changing gender mix of the profession. Unlike the other data presented in this paper a lengthy time series of data is not available on this topic. Instead I infer what I can from the cross-section of data provided by the Institute and reproduced in Table 6.

What we cannot determine directly from Table 6 is the trend of gender membership of the profession over time. What we can infer is a blueprint of the future. For example, in the over 65 age category, over 88% of the reporting members are men, while in the under 26 age category the majority of the reporting members (almost 50%) are women. In fact, women are an increasing proportion of the membership in every category as age declines. This suggests that, over time, women have and will continue to play a larger role in the profession.

This analysis is limited in several ways. There is no data available on the proportion of senior positions within the firms held by each gender and there are no compensation statistics shown here. This preliminary analysis is intended to illustrate the point that the CPA of the future is (roughly) equally likely to be male or female.

Economic Trends

I also examine the growth of the accounting profession in relation to key economic indicators in the U.S. economy. For this study, I chose Gross Domestic Product and Personal Consumption published by the Bureau of the Census. These time series are chosen, respectively, as proxies for general economic activity and consumer expenditures.[4]

As can be seen from Table 7 and Figure 4, the number of dollars of GDP per CPA and the number of dollars of Personal Consumption per CPA parallel each other closely. Both statistics appear to be levelling off somewhat over the last several decades. This suggests a growth relationship in the number of CPAs which relates that one CPA is operating for each $16.5 million of GDP or that for every $11 million in personal consumption expenditures the services of one CPA are required.

Table 7. Comparisons of AICPA Members with Key Economic Indicator $000 per AICPA member

	1930	1940	1950	1960	1970	1980	1990
GDP	149,720	164,153	86,906	50,555	38,117	23,609	16,464
Consumption	115,639	116,566	58,253	32,638	24,16	15,306	11,047

Note: [1] Personal Consumption and Gross Domestic Product number represent thousands of 1987 dollars per CPA. Both are the fourth quarter numbers of the respective year end.

The comparisons of membership with key economy-wide indicators tell a story consistent with earlier sections of this paper. The "density" of accountants in the economy has increased tremendously since 1930 but appears to be levelling off. Both of these comparisons show an increase in the relative prevalence of accountants in the economy over this century. The statistics also suggest that this relative growth is slowing, perhaps even reaching a saturation point.

CONCLUSIONS AND EXTENSIONS

The statistics presented in this paper provide four preliminary conclusions. First, both demographic and economic trends show a substantial increase in the presence of CPAs in the economy

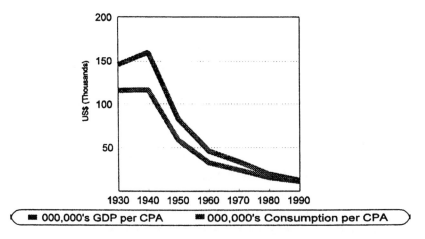

Figure 4. A Comparison of AICPA Membership with Key Economic Indicators, 1930-1990

over prior years. This result is consistent with the increasing importance of information in our society and the accounting profession's ability to provide some of that information. Second, all of the statistics presented in this paper show a slowdown in membership growth. Figure 5 clearly illustrates this point. This slowdown is occurring both in absolute growth rates and growth relative to economic indicators. Both of these conclusions suggest the mature stature and role of CPAs in the broader economy.

In contrast, the third and fourth conclusions reflect important shifts within the profession. It can be expected that, within a few years, corporate practitioners will outnumber public practitioners within the Institute. This raises a strategic policy question regarding the assurance function, the preparer function, and the role of the AICPA in each. Although a shift in each member groups' potential power is only now becoming obvious, we should anticipate these effects in the future. For example, every AICPA president since inception has been a member of the public practice community. Expectations for a change in focus and for representation are likely to grow over time. Further, the gender statistics suggest that larger proportions of younger membership are women. The necessity of accommodating these groups' needs will grow over time as corporate practitioners and women play increasingly important roles in the Institute's leadership.

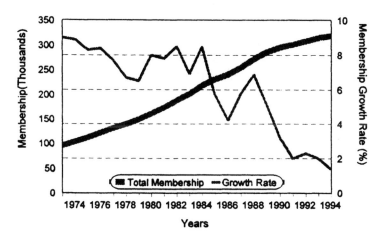

Figure 5. AICPA Membership and Membership Growth Rates 1973-1994

Two strategic challenges to be faced by the Institute's leadership in the next decade seem clear. First, how does the Institute manage slow growth in the membership? Second, how does the Institute encourage the participation and provide services to members who are not in public practice and to the increasing number of women members?

As the statistics presented in this study are preliminary and incomplete, it is not possible to provide definitive conclusions from the results developed. For example, several unanswered questions remain which related to productivity, technology and salary information. Legitimate productivity data would enable a far more in-depth analysis of the profession's contribution to society. These statistics also could be used to determine, more conclusively, the implications of slower growth in the CPA profession. Another extension would be to perform a more in-depth statistical analysis on a portion of the information presented in this study with the goal of providing statistically persuasive implications. A third extention would be to study the accounting profession in countries with similar economic and legal systems at different stages of economic development. This would provide support or refute the preliminary assertions presented here about saturation rates of professionals in a post-industrial national market economy.

In conclusion, this paper provides information on AICPA membership and its composition for an extended period of time. The data suggest that the challenges experienced in the past (i.e., rapid growth in an expanding market) will not be the challenges of the future. The ability of the Institute's leaders to adapt to the changing circumstances of the members will determine the success of the AICPA in meeting the issues of the future. Further inquiry and development of demographic and productivity profiles of the CPA profession should contribute to addressing these issues.

APPENDIX

Number of Institute Members, Actual Membership Changes, and Annual Membership Growth Rates, 1887-1994

Year	Number of Members	Change	Growth Rate (%)
1887[a]	26	26	N/A
1888	37	11	43.31
1889	32	(5)	− 13.51
1890	31	(1)	− 3.13
1891	31	0	0.00
1892	32	1	3.23
1893	55	23	71.88
1894	65	10	18.18
1895	42	(23)	− 35.38
1896	75	33	78.57
1897	89	14	18.67
1898	76	(13)	− 14.61
1899	80	4	5.26
1900	92	12	15.00
1901	112	20	21.74
1902	129	17	15.18
1903	140	11	8.53
1904	148	8	5.71
1905	587	439	296.62
1906	661	74	12.61
1907	700	39	5.90
1908	802	102	14.57
1909	873	71	8.85
1910	995	122	13.97
1911	1,093	98	9.85
1912	1,130	37	3.39
1913	1,067	(63)	− 5.58
1914	1,074	7	0.66
1915	1,058	(16)	− 1.49
1916[b]	1,238	180	17.01
1917[c]	1,220	(18)	− 1.45
1918	1,227	7	0.57
1919	1,252	25	2.04
1920	1,363	111	8.87
1921	1,484	121	8.88
1922[d]	2,259	775	52.22
1923	2,631	372	16.47
1924	3,012	381	14.48
1925	4,047	1,035	34.36
1926	4,465	418	10.33
1927	4,564	99	2.22

(continued)

APPENDIX. (Continued)

Year	Number of Members	Change	Growth Rate (%)
1928	4,660	96	2.10
1929	4,702	42	0.90
1930	4,815	113	2.40
1931	4,696	(119)	– 2.47
1932	4,405	(291)	– 6.20
1933	4,254	(151)	– 3.43
1934	4,408	154	3.62
1935	4,515	107	2.43
1936	4,835	320	7.09
1937[e]	4,890	55	1.14
1938	5,047	157	3.21
1939	5,184	137	2.71
1940	5,437	253	4.88
1941	5,722	285	5.24
1942	6,453	731	12.78
1943	7,137	684	10.60
1944	7,996	859	12.04
1945	9,051	1,055	13.19
1946	10,042	991	10.95
1947	10,954	912	9.08
1948	12,247	1,293	11.80
1949	13,960	1,713	13.99
1950	16,061	2,101	15.05
1951	17,998	1,937	12.06
1952	20,032	2,034	11.30
1953	22,038	2,006	10.01
1954	24,264	2,226	10.10
1955	26,345	2,081	8.58
1956	27,850	1,505	5.71
1957[e]	30,568	2,718	9.76
1958	32,489	1,921	6.28
1959	34,798	2,309	7.11
1960	37,897	3,099	8.91
1961	41,038	3,141	8.29
1962	44,140	3,102	7.56
1963	47,376	3,236	7.33
1964	50,413	3,037	6.41
1965	53,709	3,296	6.54
1966	57,199	3,490	6.50
1967	60,908	3,709	6.48
1968	65,115	4,207	6.91
1969	69,225	4,110	6.31
1970	75,381	6,156	8.89
1971	79,736	4,355	5.78
1972	87,562	7,826	9.81
1973	95,415	7,853	8.97

(continued)

APPENDIX. (Continued)

Year	Number of Members	Change	Growth Rate (%)
1974	103,863	8,448	8.85
1975	112,494	8,631	8.31
1976	121,947	9,453	8.40
1977	131,300	9,353	7.67
1978	140,158	8,858	6.75
1979	149,314	9,156	6.53
1980	161,319	12,005	8.04
1981	173,900	12,581	7.80
1982	188,706	14,806	8.51
1983	201,764	13,058	6.92
1984	218,855	17,091	8.47
1985	231,333	12,478	5.70
1986	240,947	9,614	4.16
1987	254,910	13,963	5.80
1988	272,479	17,569	6.89
1989	286,358	13,879	5.09
1990	295,633	9,275	3.24
1991	301,410	5,777	1.95
1992	308,280	6,870	2.28
1993	314,427	6,147	1.99
1994	318,829	4,402	1.40

Notes: [a] The American Association of Public Accountants.
[b] Institute of Accountants in the U.S.A.
[c] American Institute of Accountants.
[d] First year of data for American Society of Certified Public Accountants.
[e] Merger of the American Institute of Accountants and the American Society of CPAs on January 1, 1937.
[f] American Institute of Certified Public Accountants.
Source: AICPA Annual Reports.

ACKNOWLEDGMENT

I would like to thank Gary J. Previts, Nandini Chandar, and two anonymous referees for their helpful comments. I would also like to thank the Social Sciences and Humanities Research Council of Canada for their financial support.

NOTES

1. Other activities refer to activities pursued by members which do not fit into categories currently recorded separately by the Institute. Because I have no reliable data on subsets in this category, I do not specifically pursue it further.

2. Based on the latest five-year geometric averages, corporate practitioners will outnumber public practitioners by the 1995 reporting year. Even a narrowing of the spread between the two growth rates suggests that, by 2000, public practitioners will have less members than corporate practitioners.

3. I define the term "accountants" as CPAs and "CPAs" as members of the AICPA. Although all CPAs are not members of the AICPA, there is no reliable data available as to the quantity of these nonmembers.

4. These are admittedly rough measures. They should, however, prove to be sufficient for the purposes of this study.

REFERENCES

American Institute of Certified Public Accountants. various years. *Annual Reports*. New Jersey: AICPA.

Bureau of the Census, Department of Commerce. various years. *Census of the Population,* volume 1: *Characteristics of the Population*. Washington, DC: U.S. Government Printing Office.

Ibbotson Associates. 1993. *Stocks, Bonds, Bills, and Inflation, 1993 Yearbook*. Chicago: Author.

Pond, J.D. 1992. *Financial Advisor's Desk Reference*. New York: Simon & Schuster.

Standard and Poors Security Price Index Record, 1994 Edition. 1994. New York: McGraw-Hill.

AUDIT CONFLICT AND COST STANDARDS IN THE DEFENSE INDUSTRY

Norma C. Holter

ABSTRACT

A firm accepts accounting regulation as a part of the cost of doing business with the government. The government assures compliance with these regulations through a framework of audit and oversight. Conflicts arise during an audit, over cost determinations, and defense contractors perceive a power imbalance in favor of auditors from the Defense Contract Audit Agency (DCAA). This perception of auditor power is investigated.

INTRODUCTION

An interest to undertake this study was developed when government accounting regulations were identified as a factor thwarting the

Research in Accounting Regulation, Volume 9, pages 199-209.
Copyright © 1995 by JAI Press Inc.
ISBN: 1-55938-883-8

progress of defense contractors attempting to implement updated inventory management systems. When a contractor[1] invited the government auditors' comments on the MRP (materials requirement planning) system the firm was implementing, the issue escalated to the level of Congressional Hearings. While the inventory matter has since been resolved, the underlying cause for the notoriety— government audit for compliance with regulations—remains a contentious issue. Contractors perceive that the government auditor enjoys an imbalance of power when a conflict arises over the accounting treatment of a cost. Such an advantage may result in the inconsistent application of the Cost Accounting Standards and Principles. Also, a 1993 study concluded that compliance with government regulation cost the private sector at least $430 billion annually; audit conflict represents a portion of this cost. (*Report of the National Performance Review* 1993).

Previous researchers have studied audit conflict between the independent auditor and the client firm during an attest engagement to identify the factors that contribute to a perceived advantage enjoyed by client management. These studies were concerned with the independence of the external auditor. Nichols and Price (1976) introduced the power paradigm to accounting studies using Cartwright and Zander's (1968) definition of power as the "capability of one party to influence the attitudes or behavior of another party."[2]

This study extends previous research by examining the audit relationship in compliance auditing, focusing on a special niche of compliance auditing: the audit of Department of Defense (DOD) prime contract awards, totaling approximately $131 billion.[3] Firms manufacturing aircraft, missile/space systems, ships, and electronics and communications equipment are subject to extensive audit[4] for compliance with a myriad of Cost Accounting Standards (CAS) and Principles identified as Federal Acquisition Regulations (FAR), as well as GAAP. This study also provides the first insight into audit conflict in a regulatory environment.

THE AUDIT ENVIRONMENT

The Defense Contract Audit Agency (DCAA), a separate agency of the Department of Defense (DOD), is the primary audit presence. A DCAA audit is performed for the benefit of the government, at government expense. The objective of the audit is to provide DOD with an opinion whether the costs estimated, collected, and recorded by the contractor are fair and reasonable, and in compliance with the FAR. Conflicts arise during an audit because of a difference in interpretation (between the management of the contractor and the auditor) of some accounting practice of the contractor that may require adjustment or elimination of a cost from the price of the contract.

The complexity of accounting and costing regulations make defense contract auditing particularly difficult. A study by the Center for Strategic and International Studies found procurement regulations alone total more than 30,000 pages and are issued by 79 different offices.[5] The fact that the Cost Accounting Standards Board (CASB) has been reconvened[6] is evidence that there is concern over the present costing structure. In a report submitted to Defense Secretary Cheney (1991), the Defense Advisory Panel on Government-Industry Relations stated: "DOD audit and oversight policies and practices remain among the most contentious issues in relations between DOD and industry. Better ways must be found to balance DOD's need to ensure compliance." The key word expressed by both the industry and the government is "balance."

Banker, Cooper, and Potter (1992) suggest that in governmental auditing, a reversal occurs in the auditor-firm relationship. The government auditor is perceived as enjoying the power advantage because the government is the customer for the contractor's products, as well as the regulator of the industry. The auditor's power is also derived from the fact that payment to the firm may be withheld or the firm may be fined, suspended, or disbarred from future government contracts as the result of unfavorable audit findings.

This study employs empirical evidence to investigate the effect of five factors on the perceived power of the auditor in an audit conflict. An increased understanding of the explanatory factors of this perception should facilitate current efforts to create a more efficient government (*Report of the National Performance Review* 1993) and could help to reduce the adversarial relationship between the auditor and the contractor.

DEVELOPMENT OF RESEARCH HYPOTHESES

Earlier research (Goldman and Barlev 1974; Nichols and Price 1976; Knapp 1985; Lindsay 1989; Magee and Tseng 1990) identify the lack of specification or clarity in accounting standards as a significant factor in audit conflict. Cost Accounting Standards and Principles have sometimes been called "fuzzy."[7] The cost accounting concepts that regulate the costing of government contracts are a patchwork of rules and regulations. The procurement system has most recently been criticized as "an extraordinary example of bureaucratic red tape" (*Report of the National Performance Review* 1993, 26). Therefore, it is hypothesized that:

Hypothesis 1. Contractors perceive the auditor will win more frequently when the issue is addressed by unclear or ambiguous costing standards/principles.

The Administrative Contracting Officer (ACO) is a government employee responsible for administering the contract for the purchase of goods and services, after the award has been made, until the time of termination. The ACO is authorized to make determinations concerning contracts, with the focus on negotiating a fair and reasonable price, rather than focusing on individual elements of cost and profit. When the contractor and the auditor cannot resolve a conflict, the next recourse for the contractor is the ACO. The ACO's role could be considered similar to that of the Audit Committee. However, there is one control that differs in this environment: DOD Directive 7640.2 requires the ACO to justify deviations from DCAA recommendations. It is hypothesized that:

Hypothesis 2. The contractors perceive the auditor will win more frequently when the ACO frequently supports the auditor.

Earlier studies have utilized audit firm size to investigate the incentives to engage an external auditor, the auditor's ability to resist client pressure, and the willingness of audit committees to support auditors (Chow 1982; Knapp 1987; Lindsay 1989; Gul 1991). More audit is conducted within the large contractors, to the extent that many of these firms have "resident auditors" (government auditors who work full-time at the contractor's facility) because there is more complexity and inherent risk in their accounting systems. However, the larger contractors have the infrastructure to analyze and implement the regulations and the resources to hire knowledgeable personnel (including previous DCAA auditors) who are not intimidated by conflict with an auditor. Therefore, it is hypothesized that:

Hypothesis 3. The contractors perceive that the auditor wins more frequently when the dispute is with a small firm.

The cost-based method is the primary method of contracting for large systems and many other types of unique supplies or services purchased by the government. Cost-based contracting occurs when the price paid to the contractor is based on the estimated or actual cost incurred in producing the goods or services. Problems and conflict between the government and the contractor occur because contractors "are not made with cookie cutters."[8] Production has generally been costed by the job order, process costing, or standard costing method. This study attempts to determine if one method of costing tends to induce more audit conflict. Because process costing employs the mechanism of equivalent units to produce an average cost, it is hypothesized that:

Hypothesis 4. The contractors perceive the auditor wins more frequently when the firm utilizes the process costing method.

According to the DCAA audit manual, certain audit reports are not to be released to the contractor without the specific direction of the ACO. When a contractor requests a copy of the audit report, the auditor cites the direction of the manual, but the ACO expresses a reluctance to release a DCAA report. Some of these reports may become the basis for later claims by the government against the contractor. Possession of information is a form of power. The auditor is perceived as powerful when the contractor is not informed or unable to address some of the comments and conditions cited in an audit report. It is therefore hypothesized that:

Hypothesis 5. The contractors perceive the auditor wins more frequently when the contractor does not receive a copy of the DCAA audit report.

RESEARCH METHOD

A questionnaire was mailed to the Director of Government Contract Costing of approximately 1,000 firms, identified from a DCAA directory of firms bidding on awards totaling over $500,000 (thereby subjecting the firm to compliance with CAS). The questionnaire consisted of four parts: (1) source of conflicts, (2) conflict resolution, (3) environmental factors, such as existence of an internal audit function, a Code of Ethics, the costing method, the impact of the media, enforcement and penalties, and (4) background data about the respondent and the firm.

The directory from which the sample was drawn, in all likelihood, had attenuated by at least 35% because of the high rate of attrition of firms in government contracting,[9] the lack of diligence in keeping the list updated, and inclusion of firms that had bid for government work but did not receive an award. A follow-up mailing was sent approximately two weeks after the initial mailing. A total of 125 questionnaires were returned. Sixty-five percent of the responses came from 79 firms identified as large firms and 46 responses were from firms that regularly bid as small firms. The size of the firms responding ranged from $1 million to over $20 billion in total annual sales. Sixty-two percent of the firms were currently under audit. Eighty-three percent of the respondents had more than five years experience in the industry; 63% had more than ten years experience in the industry.

RESULTS

A multivariate linear regression model showed support for three of the research hypotheses: that the lack of clarity in the CAS and the FAR, ACO conflict resolution, and size of the contractor help to explain the perceived power of the DCAA auditor (see Table 1). The variables measuring the costing method and receipt of the audit report were not statistically significant ($p < 0.05$) although the signs of the coefficients were in the hypothesized direction. The model explained 37% of the variance.[10]

The findings of this study are consistent with earlier studies (Knapp 1985; Lindsay 1989) and contribute to the generalization of the hypothesis that the more influential party in an audit conflict gathers

Table 1. Summary of Stepwise Regression Results

Independent Variable	Standardized Beta	t-statistic	Significance of "t"
Clarity of Standards and Principles	0.3878	3.83	.0003*
ACO Conflict Resolution	0.3596	3.70	.0004*
Size of Contractor Firm	0.2501	2.50	.0151*
Costing Method	0.1175	1.19	.2381
Receipt: Audit Report	−0.1292	1.36	.1786

Notes: * Significant at $p < .05$.
$R^2 = .4154$.
R^2 adjusted $= .3711$.
Regression F-value: 9.38 (significance = .00001).
Durbin-Watson: 2.25.
Residual diagnostic: Kolmogorov-Smirnov Z = .648 (p-value .795).

some of its perceived power from unclear standards. This study also confirms that size is a significant factor (Lindsay 1989).

DISCUSSION

The CAS and the FAR were written with a degree of ambiguity so they could be applied across diverse firms producing many different products for the government. The findings indicate that this ambiguity increases conflict and works to the advantage of the auditor. The FAR may not be consistently applied, by either the contractor or the auditor, resulting in some distortion in the costing process. One respondent commented: There is conflict over "disallowance of economically necessary and fair expenses just because the auditor thinks that the regulations or some tradition decree that such expenses are not allowed." Interestingly, the *Report of the National Performance Review (NPR)* cites incidents of inefficiency in the procurement process because of "powerful tradition." Another respondent felt that: "The elements of cost comprising G & A [general and administrative expenses] are in many ways misunderstood by DCAA as to their need, why they are required in running a business." Again, in discussing the audit of G & A, one respondent commented that "the DCAA has as many versions as they have auditors." Many of the other comments can be summarized in this statement by a small contractor: "There are too many standards now. There should be a broad default to GAAP for most instances."

One respondent, a former DCAA auditor now working for a contractor, offered the following insight.

> When I was with DCAA we generally interpreted the CAS differently than the contractors. The DCAA training focused on the theory behind CAS rather than the practical application of the standards. Examples of how certain contractors properly applied CAS were not given. Also, the DCAA did not have a universal concept of materiality which led to much wasted time pursuing CAS issues with little or no savings to the Government.... The Government brings numerous problems upon itself by attempting to micro-manage the contractors accounting through the CAS and the FAR.

The National Performance Review (NPR), under the direction of Vice President Al Gore, has recommended that the procurement process be simplified by rewriting federal regulations (the FAR and agency supplements) shifting to guiding principles (*Report of the National Performance Review* 1993, 28). The new regulations are to end unnecessary regulatory requirements. The NPR should coordinate with the CASB. Comments should be solicited from the government auditors and the contractors regarding the weak, as well as the effective sections of the current CAS and FAR. Clarification or elimination of all weak, ambiguous, contentious areas should be considered in this effort.

If the contractor believes the ACO favors the auditor in the resolution of a conflict, the contractor will attempt to resolve the dispute with the auditor by compromise and negotiation. This practice, exacerbated by ambiguous standards, could lead to possible distortion and game playing in the costing process. This finding confirms the conclusions of the *Packard Report* (1986) and leaves the same issue unresolved. Is the ACO agreeing with the auditor because the ACO believes the auditor is a better authority on interpreting the FAR? Or, is the ACO agreeing with the auditor in order to avoid a confrontation under Directive 7640.2? A review of Directive 7640.2 and a study of its implications on the ACO/auditor relationship should be considered. Nash and Cibinic (1993) summarized the situation as follows: "Many (A)COs have capitulated to the auditors in matters related to accounting and pricing. The (Audit) Handbook clearly indicates that the auditor has an important function to perform and that the (A)CO must utilize the auditor's services when appropriate. However, it also makes it clear that the (A)CO is not to play second fiddle to the auditor."

The findings confirm the auditor wins more frequently in a conflict with a small firm. The comment by one respondent, "we have no choice but to accept the auditor's decisions" summarizes the dilemma seemingly faced by the smaller firms. The number of smaller firms will continue to increase because more work is being delegated to subcontractors. The dynamics of fabricating the product have changed; the product is now assembled by the prime contractor from components supplied by the subcontractors. The NPR revision of the FAR presents the perfect opportunity to write costing guidelines that could be applied by all firms, regardless of size or industry. This is not an easy task. However, the fact that most of the conflict develops in the gray areas of accounting systems indicates that a substantial portion of the present costing structure works. The NPR is confident that the recommended reengineering of government can be accomplished to save a projected $108 billion over five years.[11]

A majority of the respondents indicated use of the job order costing method. Many firms, including the smaller contractors, utilize purchased software to collect the data. This could be construed that the hypothesis concerning process costing is not relevant. Costing problems seem to focus on the allowability of various factors of overhead and general and administrative expense. One respondent expressed concern that "the DCAA wants us to change our accounting systems to satisfy their reports."

Contractors get a copy of the audit report eventually, although as one respondent commented, "it's like pulling teeth." This situation should be addressed by the DOD. Communication to the contractors of audit results could ease tensions, provide valuable feedback to the contractors, and perhaps reduce litigation in the contract area. Several contractors mentioned that the present DCAA reward system encourages an adversarial relationship between the auditor and the contractor. A comment by one respondent that "as long as DCAA promotion is based on costs recovered, there will be conflicts" should be addressed by the NPR, specifically the interagency regulatory coordinating group. More positive incentives for promotion and recognition should be developed.

Audit and oversight for compliance with ever-increasing laws and regulations is becoming more extensive and expensive for the

Government and the private sector, to the point where many firms are questioning the cost of doing business with the government. This study examined one niche of compliance auditing—the audit of Department of Defense contracts.

NOTES

1.   Government regulations and contract accounting literature describe firms that contract to provide the U.S. Government, particularly the Department of Defense, with goods and services, as defense "contractors."

2.   Nichols and Price also incorporate Emerson's (1962) theory of power relations utilizing Emerson's power-dependency equation, that is, Pab = Dba (power of A over B rests implicitly in B's dependence on A) and Pba = Dab (the power of B over A is equal to A's dependence on B.)

3.   *The Government Contractor*, "DOD Issues FY 1993 Report on Prime Contract Awards by State" (Vol. 36, No. 13, March 30, 1994, p. 8).

4.   Nash and Cibinic, *The Nash & Cibinic Report*, "Dateline July 1993" (Vol. 7, No. 7, July 1993). "The government audits and regulates its contracts to a degree unknown in the rest of the world—presumably to demonstrate to the public that it should trust the procurement process."

5.   This study, quoted in Monahan and Claiborne (1988), also noted that defense activities are monitored by 55 subcommittees of 29 congressional committees.

6.   The initial Cost Accounting Standards Board (CASB) was legislated into existence in 1970, with a mission to achieve increased conformity in accounting practices among government contractors and greater consistency in the accounting treatment of costs. It was quietly extinguished during the spate of deregulation in the early 1980s. The CASB was re-established under Public Law 100-679 on November 17, 1988.

7.   A description of the Cost Accounting Standards and Principles in testimony before the U.S. Congress, House Armed Services Committee, Subcommittee on Readiness, 99th Congress, 2nd Session, June 1987.

8.   Nash and Cibinic, *The Nash & Cibinic Report*, "Dateline April 1994" (Vol. 8, No. 4, April 1994).

9   Beighle (1990) comments: "From the statistics I have seen, over half of the firms that were part of the defense industrial base have disappeared from government contracting over the past three to four years."

10.   By comparison the use of multivariate regression analysis in this study has produced a greater percent of explained variance than has been achieved in prior studies of audit conflict utilizing ANOVA: Shockley (1981), 28% for Big Eight partners, 19% for other CPA partners, 23% for commercial loan officers, and 24% for financial analysts; Knapp (1985), 32%; Lindsay (1989), 12%; and Gul (1991), 17%.

11.   See *Report of the National Performance Review* (1993, Appendix A and B). Savings is projected over a five-year period, FY 1995-1999.

REFERENCES

Banker, R.D., W.W. Cooper, and G. Potter. 1992. A perspective on research in governmental accounting. *The Accounting Review* 67(July): 496-510.

Beighle, D.P. 1990. Defense contractors—the next spotted owl? *National Contract Management Journal* 24(1): 23-30.

Cartwright D., and A. Zander. 1968. *Group Dynamics: Research and Theory.* New York: Harper and Row.

Federal Contracts Reports. 1991. Washington, DC: Bureau of National Affairs, March 25, pp. 400-414.

Goldman, A., and B. Barlev. 1974. The auditor-firm conflict of interests: Its implications for independence. *The Accounting Review* (October): 707-718.

Gul, F.A. 1991. Size of audit fees and perceptions of auditors' ability to resist management pressure in audit conflict situations. *Abacus* 27(2): 162-169.

Knapp, M.C. 1985. Audit conflict: An empirical study of the perceived ability of auditors to resist management pressure. *The Accounting Review* 60(April): 202-211.

Lindsay, D. 1989. Financial statement users' perceptions of factors affecting the ability of auditors to resist client pressure in a conflict situation. *Accounting and Finance* 29(November): 1-18.

Monahan, T.F., and C. J. Claiborne. 1988. Self-policing strategies for defense contractors. *Internal Auditor* (December): 17-21.

The Nash & Cibinic Report. 1989. The government's status as a contracting party. Washington, DC, March, pp. 41-42.

Nichols, D.R., and K.H. Price. 1976. The auditor-firm conflict: An analysis using concepts of exchange theory. *The Accounting Review* (April): 335-346.

The Packard Commission. 1986. *A Quest for Excellence.* Final Report to the President by the President's Blue Ribbon Commission on Defense Management, a.k.a. The Packard Report, June.

Report of the National Performance Review. 1993. From red tape to results: Creating a government that works better & costs less by Vice President Al Gore. Washington, DC: U.S. Government Printing Office, September 7.

Shockley, R.A. 1981. Perceptions of auditors' independence: An empirical analysis. *The Accounting Review* 56(October): 785-800.

U.S. Defense Advisory Panel on Government-Industry Relations. 1991. Report to U.S. Defense Secretary Cheney. In Confronting the Realities of Government Audit Practices. Survey by Manufacturers' Alliance for Productivity and Innovation, August. p. 1.

U.S. Department of Defense. 1982. Directive 7640.2, December 29.

―――. 1993. Defense Contract Audit Agency. *DCAA Audit Manual* (December): 101, 601.

ASSESSING THE UTILITY OF CONTINUING PROFESSIONAL EDUCATION FOR CERTIFIED PUBLIC ACCOUNTANTS

Paul J. Streer, Ronald L. Clark, and Margaret E. Holt

ABSTRACT

Over the past decade there has been significant growth in continuing education for CPAs. There is, however, little research that measures the effectiveness of the profession's CPE efforts. Reports by the Sanford and Netterville committees on the AICPA's role in CPE suggest the need for additional research. We conducted a survey of accountants to determine the impact of CPE courses on their practice. Adult and continuing professional education literature suggests that "transfer of learning" is an important element of CPE effectiveness. Some indicators of learning transfer are use of materials, follow-up on points raised during the course, and discussion of course content with colleagues back in the workplace. Our findings have several implications for regulators and providers of CPE programs.

Research in Accounting Regulation, Volume 9, pages 211-222.
Copyright © 1995 by JAI Press Inc.
All rights of reproduction in any form reserved.
ISBN: 1-55938-883-8

Learning is an investment. If the learners apply back at work what they acquired during workshops, seminars, and other professional development activities, there will be a return on the investment. If they do not, then the training time was merely spent (and hence wasted) rather than invested.

Scott Parry (1990)

Those interested in continuing professional education generally and providers of continuing education for accountants specifically should consider the effectiveness of transferring learning from courses to work settings. Parry (1990) proposed that personal, instructional, and organizational factors "help" or "hinder" transfer of learning. These factors are pertinent to continuing professional education for CPAs.

A personal factor: *relevance* (Does the learner see the course as relevant to the job and to personal needs?)

An instructional factor: *emphasis* (Theory vs. practice? Knowledge vs. skills? Talking vs. doing?)

An organizational factor: *degree of fit* (Do local procedures, forms, and equipment agree with those taught to the learner?)

Today, most State Boards of Accountancy have adopted mandatory CPE requirements for licensing. According to the National Association of State Boards of Accountancy (NASBA), only 2 of the 54 jurisdictions currently do not have mandatory CPE (NASBA 1992). Recently, the American Institute of Certified Public Accountants (AICPA) specified annual CPE as a requirement to maintain membership.

The concern over continuing education for CPAs relates to, as Parry suggests, a positive cost/benefit to the practitioner. CPAs invest significant resources (time and money) into CPE each year. They demand that these "courses" are *relevant* to current professional needs. CPAs also expect the course to have an appropriate level of *emphasis* on theory and practice, and that it *fits* within the parameters of their organization.

A primary justification for mandatory CPE is the belief that CPAs have an obligation to maintain acceptable levels of professional competence and that CPE is an effective tool for discharging this obligation (AICPA 1986). Support for the conclusion is found in the results of a multiyear study on the effectiveness of a three-year mandatory CPE program for practitioners in New York state (Grotelueschen Associates, 1990). The study concluded there is a positive relationship between a practitioner's participation in CPE courses for a prior twelve-month period and the practitioner's knowledge proficiency as measured by a test score. In addition, a longitudinal analysis of the data found an improvement in the accountant's knowledge over a two-year period during which the mandatory CPE requirements were in effect.

Receiving CPE credit, however, is not by itself a measure of competence nor does it monitor the application of any knowledge acquired (AICPA 1980). CPE must achieve two important objectives if it is to improve professional competence. First, the participants must learn the relevant information presented to them in the courses they attend. Second, they must apply this knowledge in their respective practice environments (Knox 1979). The Armstrong Committee (AICPA 1980) suggests there is a near universal acceptance of the notion that the typical CPA learns worthwhile information in CPE. There is, however, a current debate over the effectiveness of applying this new knowledge to practice.

Two recent reports highlight the concerns over our present CPE delivery model. The *Report of the Special Committee on the Future of CPA Continuing Professional Education* (AICPA 1993), recommended creating an "Alliance for Learning" to coordinate and distribute CPE materials. One of the alliance's duties would be to "undertake a study of its full CPE product line, most notably the group study programs" (p. 15). The Sanford Committee (AICPA 1993) also called for more technology based and innovative methods of delivering CPE.

The AICPA's *Report of the Task Force to Evaluate the Recommendations Contained in the Sanford Report* (AICPA 1994) did not agree on the need for creating an "Alliance for Learning." The task force noted that "recent CPE surveys have provided strong evidence that CPAs still prefer group study seminars and conferences" (p. 19).

These reports and those previously cited bring into question the utility of present methods of providing CPE for the accounting profession. This paper describes an investigation to learn the extent to which CPE participants used program information in their practices within a specified post-course interval. The results of this study should prove useful to State Boards of Accountancy and the AICPA in formulating more precise and effective CPE requirements for licensing and membership purposes. Other professions concerned about the impact of their continuing education programs and requirements may find our results helpful. Also, adult educators who investigate practice and theory in continuing professional education may find the results valuable. Hopefully, our study will also provide useful input into the current debate surrounding CPE delivery.

DESCRIPTION OF STUDY

Although learning theory in adult education is evolving, new research in the area of situated cognition offers significant support for more effective learning transfer if learning can be applied to real contexts and situations (see, for example, Rogoff 1984; Schon 1983). Situated cognition research suggests that learning is not complete until one applies the new knowledge. For accountants, the acquisition of knowledge gained through CPE courses should influence their work in subsequent periods. In other words, CPAs should find actual uses, while performing their professional duties, for the specific information contained in the courses they attend. A CPA could experience an increase in his/her efficiency or effectiveness because of a particular point covered in a course. On the other hand, the impact of the course might be more general in nature. The course could create a positive influence on the accountant's attitude or approach to doing his/her work. Whatever the type of impact, for CPE to effectively enhance professional performance, it must significantly influence the CPA. In the current study, we undertook an investigation to determine if CPAs experienced any direct or indirect impact from participating in CPE courses.

For some time now, adult learning researchers have seen the need for impact evaluations of CPE courses (Knox 1979; Grabowski 1983; Holt and Courtenay 1985). They suggest that there should be a relationship between taking a specific course and subsequent changes in behavior. Travers (1977) uses the term "transfer of training" to indicate the successful facilitation of learning in new situations. Holt and Courtenay (1985, 23) point out, however, that "unfortunately, there is considerable evidence in recent literature that learning transfer from program to post-program settings often fails."

Cervero (1988) argues that an acceptable course-evaluation approach is to test for the application of learning after the CPE program. His work addresses the degree to which knowledge, skills, and attitudes learned from a particular program impact a relevant work setting. Other studies such as Mays (1984), Poilet and Hungler (1985), and Long and Fransen (1986) have successfully used a self-reporting method to assess this application of knowledge. Poilet and Hungler (p. 182) state that "perhaps the strongest argument that can be made about the self-reporting method is that it frequently yields information that would be difficult, if not impossible, to gain by any other means." We also make use of this self-reporting methodology in the present investigation.

We selected 14 courses offered by a state CPA society during October through December of 1991 for inclusion in the study. Time and resource constraints imposed by the state CPA society influenced the number of courses and participants we could study. Selecting 14 courses provided a reasonably sufficient sample size while keeping the workload on the state society staff to an acceptable level. The fall season is a time of peak demand for state society courses because the workload for many accountants is relatively low during this time of the year and the reporting period for both state licensure and AICPA membership purposes ends on December 31. To ensure diversity of course content, 7 of the courses chosen were tax related. The remaining 7 were financial accounting or auditing courses. We also obtained personal and employment data for each attendee in each course selected.

A questionnaire was developed based on the adult education research of Bryant and Shinn (1979), Caffarella (1994), and Phillips (1994). Their research of effectively analyzing and capturing the degree of post-course learning application provided a basis for the survey. For example, Bryant and Shinn provide several examples of how a CPE participant could successfully apply new knowledge to reinforce learning, and experience a direct impact on his/her performance. Using this prior research base as a backdrop, we selected the following four closed-end questions for inclusion in our survey questionnaire.

Survey Questions

1. Did the information presented in this course have an impact on you and/or your firm or company?

2. Have you used the course material following the course?
3. Have you followed up on any particular material you obtained in the course?
4. Did you discuss any of the information from the course with other staff or colleagues following the course?

Following the suggestions of Bryant and Shinn, Questions 2, 3, and 4 determine how the CPA might have applied the material to reinforce specific technical content or general concepts learned in the CPE course. Question 1 measures the participant's perception of the impact of the course on his/her practice. We also obtained some information regarding the participant's motivation for taking the particular course under investigation. The available responses were: (1) employer requirement, (2) need CPE for licensure, (3) relevance to business, (4) timeliness of topic, and (5) all other reasons—specify. While there were some responses to each category, the vast majority of participants indicated that they took the particular course to satisfy licensure requirements.

The questionnaire was pre-tested using 6 randomly selected participants. No significant revisions to the questionnaire content were made based on an analysis of their responses. There was a total of 373 course participants in the 14 courses selected. The participants were evenly divided between those in public practice and those in industry. Each participant received a questionnaire, accompanied by a cover letter signed by the CPE director. The letter urged the participants to complete the questionnaire and indicated that their willingness to do so could benefit both them and the state society by leading to improved CPE course offerings. The mailing also included a brief general description and major topic outline for the course they had taken. These items helped ensure accurate responses based on the specific CPE course under investigation.

For several reasons, we used a 6- to 9-month time span between course participation and data collection. It gave the subjects an opportunity to use the course information in their work environments. The period also encompasses one accounting busy season for those subjects in public accounting practice. In addition, we believed that the participants would be much more likely to complete the questionnaire when their workload was not at its peak. Furthermore, when investigating learning application, the adult education literature does not reveal any empirically preferred interval between CPE exposure and data collection.

Two weeks after mailing the initial questionnaire, the CPE staff of the state society called each nonrespondent. We received 153 usable responses which represent an overall response rate of 41%. Eighty-six percent of the respondents were residents of the sponsoring state while the remaining respondents were residents in ten different states. These demographics give the study geographic dispersion and provide generalizability of the research findings and their consequences to the entire population of U.S. CPAs.

RESULTS OF THE STUDY

Table 1 presents a summary of the responses to the four research questions. The table clearly shows that most of the respondents attributed some impact on their practice to the specific CPE course they took. What is not clear, however, is the relationship between IMPACT and the three variables measuring application of the new knowledge. We used chi-square analysis to test the relationship of the course's IMPACT on the respondent's practice and use of materials (USE), follow-up on information received in the course (FOLLOW), and discussion of material with other colleagues (DISCUSS).

IMPACT relates positively to USE of materials and FOLLOW-up on information received from the CPE course. To a lesser degree, DISCUSS also relates to IMPACT and reinforces the idea that sharing new information places greater importance on the items learned (chi-square and P-values for USE, FOLLOW, and DISCUSS were 23.64, .0000; 11.02, .0009; and 5.80, .0161, respectively).

Our findings support learning theory that suggests CPE effectiveness relates to the application of new knowledge. These findings are consistent with a study by Courtenay and Holt (1990, 14) that measured course impact on professionals who attended a symposium on understanding and managing human relations. In that study, the authors report that the majority of participants (75%) received some impact from the course. The authors also found that 62% followed up on points raised during the symposium, 82% used the information in their jobs, and almost all discussed the topic further with colleagues.

From the current study, however, it is important to note that 31% of the respondents indicated they received **no** impact from the course.

Table 1.   Summary of Findings

Variable	Yes	No
Impact on Practice (IMPACT)	105	48
Use of Materials (USE)	75	78
Follow up (FOLLOW)	26	127
Discuss material (DISCUSS)	73	80

There could be many reasons why these individual respondents perceived the course to have no influence. Our data allow us to test the impact of two possible explanatory variables. CPE participants can be grouped depending on whether or not they practice public accounting. Those not in public practice generally are either in industry or government accounting positions. Many accounting CPE courses stress topics which appear more directly related to public practice. These courses cover such items as gathering audit evidence, compilation procedures, and preparation of individual tax returns. One might argue, therefore, that those participants not in public practice would find it more difficult to find a relationship between a CPE course and their particular job. Of the 48 respondents who indicated no impact, 33 were not in public practice. Only 9 of these respondents used the materials and only one followed up on course comments and points. It is somewhat surprising, however, that 16 of the 33 discussed the course with their colleagues. We analyzed the survey data to determine if area of practice influenced the perceived usefulness of the courses. Again, we used chi-square to determine if there was a relationship between practice area and usefulness of course.

Our tests indicate that those respondents in public practice received a higher impact from the courses. These respondents use the material more, and follow-up on course points more than those not in public practice (chi-square and p-values for IMPACT, USE, FOLLOW, and DISCUSS were 7.02, .008; 7.96, .005; 4.22, .034; and .04, .834, respectively).

We also tested the data by looking at length of experience. Those respondents with ten years or less experience (74) and those with over ten years of experience (79) form the two groups. The results suggest there were no significant differences between the two groups and, therefore, experience had no influence on course impact (the chi-square and p-values were 1.71, .187; .01, .929; 3.88, .049; and .56, .455).

IMPLICATIONS OF STUDY

Participants acquiring new knowledge and skills which are applied to their relevant occupation is one crucial outcome of any CPE program. Our findings support this notion and are consistent with learning theory objectives and other studies in continuing education. The results of our tests suggest that, to complete learning, knowledge must "transfer" from the course environment to the practice environment. This phenomenon is not unique to accounting CPE. For example, an earlier study involving medical education found that sometimes participants did not retain their improved performance six months after taking a CPE course (Knox 1979). Using the material, following up on course points, and discussing course contents are three ways the adult learner can ensure some relevant impact from CPE courses.

Numerous factors mitigate against transfer and application of what was learned once a professional leaves the learning environment. Individuals need strong motives and incentives to use the new knowledge and skills obtained in a CPE course on the job (Grabowski 1983). Mandatory changes in accounting standards and income tax laws undoubtedly give some positive encouragement to CPAs to apply the knowledge obtained in CPE courses. Significant problems, however, still exist as the results of this study clearly show.

Furthermore, it normally takes a significant amount of time for the student to integrate new learning. Individuals participating in CPE change although the outcome is often not the behavior change intended by the program. Instead, it is normally a small step toward ultimate change. Thus, viewing CPE as a change process rather than as a series of discreet and independent learning activities is more productive (Best et al. 1989).

We believe that the motivational and process views discussed are valid. These ideas have several important implications for the regulators and providers of accounting CPE in their ongoing efforts to improve course effectiveness. As discussed in the Netterville Report, the AICPA is considering ways to develop a national CPE curriculum to enhance the competence of participants in its CPE programs. One of the AICPA's proposed projects calls for classifying CPE course content into learning units and constructing logical sequences of courses. By linking content, courses can build on and reinforce concepts and skills.

Participants are then encouraged to take courses sequentially within a given subject area of their choosing rather than in a random, haphazard fashion. We believe this program should be completed and instituted because it represents an efficient and effective means of encouraging the actual implementation of the knowledge and skills acquired in the CPE courses attended. In addition, State Boards of Accountancy could accomplish similar ends by mandating that licensees prepare an annual integrated program of study that would encompass the next licensing period. The program could serve as a flexible benchmark and coordinate a planned building block approach to knowledge acquisition within specified areas of study. Such a program could lead to an enhanced application of the knowledge obtained from CPE material by licensees in their practice environments. It could also improve the CPA's course selection procedures. Perhaps with such a program, educational considerations would dominate the need to obtain credit hours.

Sixty-nine percent of the respondents in our study said they received some impact from their CPE course. While this high success rate may be acceptable, it is significant to note that the same percentage (69%) of those responding no to course impact were not in public practice. It is vital, therefore, that providers of CPE develop methods, courses, and programs that help participants successfully complete the learning process regardless of their area of practice. Our findings suggest that CPE content for participants in industry should be differentiated from those in public practice. State Societies of CPAs need to understand this market difference and State Boards of Accountancy should begin to explore CPE rules which factor in content/practice differences.

Thoughtful focus on learning transfer techniques exists in the training, and continuing professional and adult education literature (Parry 1990; Swanson and Nijhof 1994; Fox 1994; Sleezer 1994; Nolan 1994; Cheek and Campbell 1994). Clearly, individuals who design and implement continuing professional education have a fiduciary responsibility to the consumers of professional expertise to evaluate the effectiveness and utility of legally mandated continuing education.

REFERENCES

American Institute of Certified Public Accountants. 1980. Armstrong Committee Report. *Report of the Special Committee on Regulation of the Profession.* New York: AICPA.

————. 1986. Anderson Committee Report. *Restructing Professional Standards to Achieve Professional Excellence in a Changing Environment.* New York: AICPA.

————. 1993. Sanford Committee Report. *Report of the Special Committee on the Future of CPA Continuing Professional Education.* New York: AICPA.

————. 1994. Netterville Committee Report. *Report of the Task Force to Evaluate the Recommendations Contained in the Sanford Report.* New York: AICPA.

Best, J.A., K.S. Brown, R. Cameron, E.A. Smith, and M. MacDonald. 1989. Conceptualizing outcomes for health promotion programs. In *Evaluating Health Promotion Programs,* Vol. 43, ed. M.J. Breaverman, 33-45. San Francisco: Jossey-Bass.

Bryant, B., and E. Shinn. 1979. *Follow-up Conference Assessments Impact Study (Office of Education, DHEW).* Washington, DC: Teachers Corps.

Caffarella, R. 1994. *Program Planning for Adults: A Comprehensive Guide for Adult Educators, Trainers, and Staff Developers.* San Francisco: Jossey-Bass.

Cervero, R.M. 1988. *Effective Continuing Education for Professionals.* San Francisco: Jossey-Bass.

Cheek, G.D., and C. Campbell. 1994. Help them use what they learn. *Adult Learning* (March/April): 27-28.

Courtenay, B.C., and M.E. Holt. 1990. Using impact evaluations to improve marketing plans in continuing higher education. *Continuing Higher Education* (Winter): 10-15.

Fox, R.D. 1994. Planning Continuing Education to foster transfer of learning. *Adult Learning* (March/April): 24-25.

Grabowski, S.M. ed. 1983. How educators and trainers can ensure on-the-job performance. In *Strengthening Connections Between Education and Performance,* Vol. 18, 59-72. San Francisco: Jossey-Bass.

Grotelueschen Associates, Inc. 1990. *Final Technical Report: The Effectiveness of Mandatory Continuing Education for Licensed Accountants in Public Practice in the State of New York.* New York: Author.

Holt, M.E., and B.C. Courtenay. 1985. An examination of impact evaluations. *Continuum* (Winter): 23-35.

Knox, A.B. 1979. What difference does it make? *New Directions for Continuing Education,* Vol. 3, 1-28. San Francisco: Jossey-Bass.

Long, J., and S. Fransen. 1986. *Evaluating the Trustworthiness of Self-Assessments.* ERIC Document Reproduction Service No. ED274 878. Paper presented to the American Evaluation Association, Kansas City.

Mays, M.J. 1984. Assessing the change of practice by physical therapists after a continuing education program. *Physical Therapy* 64: 50-54.

National Association of State Boards of Accountancy (NASBA). 1992. *Summary of State Continuing Professional Education Rules and Regulations.* New York: NASBA.

Nolan, R.E. 1994. From the classroom to the real world. *Adult Learning* (March/April): 26.

Parry, S. 1990. Ideas for improving transfer of training. *Adult Learning* 7: 19-23.

Phillips, L. 1994. *The Continuing Education Guide: The CEU and Other Professional Development Criteria.* Dubuque, IA: Kendall Hunt Publishing.

Poilet, D., and B. Hungler. 1985. *Essentials of Nursing Research: Methods and Applications.* Philadelphia: J.B. Lippincott.

Rogoff, B. 1984. Thinking and learning in social context. In *Everyday Cognition: Its Development in Social Context Editors,* eds. B. Rogoff and J. Lave, 1-8. Boston: Harvard University Press.

Schon, D. 1983. *The Reflective Practitioner: How Professionals Think in Practice.* New York: Basic Books.

Sleezer, C.M. 1994. Transfer analysis: fitting learning to the context. *Adult Learning* (March/April): 25-26.

Swanson, R., and W. Nijhof. 1994. Measuring transfer of learning. *Adult Learning* (March/April): 28-29.

Travers, R.M.W. 1977. *Essentials of Learning.* New York: Macmillan.

PART IV

BOOK REVIEWS

Setting Standards for Financial Reporting:
FASB and the Struggle for Control of a Critical Process

by Robert Van Riper
(Westport, CT: Quorum Books, 1994; $49.95, 216 pp.)

Reviewed by **Elliott L. Slocum**

Robert Van Riper's book regarding the accounting standard-setting process should cause some serious reflection in the accounting profession. Although some history of events prior to 1973 is included, he focuses on the Financial Accounting Standards Board (FASB). As a senior member of the staff of the FASB, 1973-1991, and with prior business experience as senior vice president of N. W. Ayer, Inc., Van Riper's observations and concerns are based on substantive experience. The book is well-documented, and the footnotes and bibliographic sources represent a significant database for further study of the accounting standard-setting process.

The reader is provided much information about the standard-setting process and the debate surrounding the objectives of accounting standards. Relationships of the FASB with the Securities and Exchange Commission and with other constituencies are an important part of the book. However, more than half of the book is dedicated to Van Riper's major concern: that corporate America has politicized the standard setting process, and as a result, the future of standard-setting in the private sector is in question.

Research in Accounting Regulation, Volume 9, pages 225-227.
Copyright © 1995 by JAI Press Inc.
All rights of reproduction in any form reserved.
ISBN: 1-55938-883-8

According to Van Riper, most participants in the standard-setting process "do not yet acknowledge that a fundamental philosophical debate is under way." However, all have some awareness that a powerful ideological argument has entered the debate. Should the FASB focus on reliability and objectivity of financial information, which follows the thrust of the Securities Acts to protect investors and to provide them adequate information to make "economic decisions"? Or, should the FASB focus on the potential economic and social consequences of reported financial information, which supports pragmatism or practice-based standards? Since 1980, the debate has led to an intense confrontation with corporate America over control of the process and renewed questions of the future of standard setting in the private sector.

Debate regarding the objectives of financial reporting and a survey in 1980 show that only a few corporate executives accept "fulfillment of user need" as the primary purpose of financial reporting. Prominent corporate executives have taken the position that the system of financial reporting is designed to serve management's needs, and that management should determine what information external users should have. Corporate America's view of the objectives of financial reporting supports the economic and social consequences or practice-based approach to standard setting.

Van Riper responds to published comments, positions, and actions of various individuals representing corporate America and the numerous complaints and criticisms of the FASB, its technical staff, and the standard-setting process. Failing to influence the FASB to directly negotiate the content of standards issued in the 1980s, members of the Business Roundtable and the Financial Executives Institute actively campaigned to have the "right" people elected as trustees and appointed to the FASB and review committees. Van Riper cites this and other instances where corporate America has attempted to increase its influence and politicalization of the process. Under increasing criticism and pressure from the business community, the trustees have tried to appease the critics of the standard-setting process. The recently elected vice president of the Foundation stated that the FASB exceeded its proper mission: to merely identify and catalog "accepted practices."

In 1990, corporate America successfully campaigned to reinstate the supermajority vote rule which clearly was intended to make change more difficult and to reemphasize consensus or the general acceptance idea of standard setting. Further support for the economic and social consequences approach was gained with the appointment of a new SEC chairman who believes that accounting rules are too complicated and theoretical and have impeded American business' ability to compete in world markets. Other changes in the business and professional environment during the late 1980s, such as federal deregulation and competition among CPA firms, have also increased pressure on the FASB.

Van Riper believes that the Foundation trustees have shown an inability or unwillingness in recent years to give reasoned, principled support to the standard-setting structure and process. Failure to recognize that an independent, neutral approach to standard setting is required to achieve public interest, portends collapse of standard setting by the private sector before the end of this decade. He suggests that the qualifications and basis on which trustees are chosen needs to be changed.

Van Riper concludes that public interest depends on meaningful standards of financial accounting and reporting based on technical validity which can only be achieved through a neutral process. Imposing the economic and social consequences approach will place control of standard setting in the hands of business interests, the politics of which will favor only a few. Worse than the political ramifications, standards derived from the negotiated, case-by-case, ad hoc standard-setting approach will lead to confusion and inconsistent and contradictory standards which would fail to satisfy the requirements of the SEC and Congress. Corporate America's lack of appreciation of the consequences of such a politicized and fragmented approach to standard setting is alarming. Failure of the private sector to set standards which recognize public interest will lead to alternatives which are almost certain to be unattractive to corporate America and all other constituencies.

Accounting Certification, Educational, & Reciprocity Requirements: An International Guide

by Jack Fay
(Westport, CT: Quorum Books, 1992; $65.00, 320 pp.)

Reviewed by **Tonya K. Flesher**

The purpose of this book is to provide information about requirements to become certified as a professional accountant and about reciprocity, continuous education requirements, and accounting organizations in various countries throughout the world. The author gathered the data on these requirements by sending questionnaires to officials in different countries in 1987 and 1991. Officials from only 42 countries responded. Information on other countries was obtained from a variety of sources.

The chapters in the book, after the introduction, are titled as follows: Reciprocity for Professional Accountants; Certification Requirements; Activities and Responsibilities of Professional Accountants; Continuous Education Requirements; Accounting Organizations and Journals; Ethics; and Recent Developments and Conclusion. There is an appendix that contains the names, addresses, and telephone numbers of over 200 accounting organizations and standard-setting bodies from over 100 countries. There is a three-page bibliography. Finally, there is an index. The only main entries in the index are for countries except two (questionnaires and international organizations).

Less than 60% of the book is textual; the remainder consists of tables listing requirements or organizations by country. Most of the narrative is descriptive of the data in the tables. The final chapter devotes a couple of pages to commentary concerning developments in the former Soviet Union.

Research in Accounting Regulation, Volume 9, pages 229-230.
Copyright © 1995 by JAI Press Inc.
ISBN: 1-55938-883-8

There are a few errors and omissions. (I could only check the accuracy of the data pertaining to the United States.) The name of the firm, Deloitte & Touche, was consistently misspelled. At least two organizations were omitted—The Academy of Accounting Historians (an international organization) and the Information Systems, Audit and Control Association (formerly the EDP Auditors Association).

As the author admits, this type of information is constantly changing. He concludes the book with a request for readers to submit corrections, additions, or suggestions for future editions. The author may want to consider using a looseleaf format for future editions so that updated material could easily be added. If this is not possible, perhaps a paperback version might make the book less expensive.

The author is to be commended for this worthwhile endeavor. This is a valuable reference book. Readers may consult this study for an overview of the requirements for many countries and can use the addresses provided to write for additional and current information for a country.

The Continental Bank
Journal of Applied Corporate Finance

(New York: Stern Stewart Management Services, Inc., 1993; Vol. 5, No. 4)

Reviewed by **Nandini Chandar**

This issue of the *Journal of Applied Corporate Finance* focuses on the economic role played by the U.S. Securities and Exchange Commission, as the principal regulator of the American securities markets. It consists of 12 out of 39 papers by distinguished economic and legal scholars on different aspects of securities markets regulation including disclosure requirements, accounting standards, corporate governance, and market structure issues that were raised by SEC's "Market 2000" study. Also included is a roundtable discussion of the recent SEC-mandated disclosures on executive pay. The complete set of papers is published by Business One Irwin as "Modernizing U.S. Securities Regulation." The research was commissioned by Kenneth Lehn, the Chief Economist of the SEC from 1987 to 1991, and now the first Director of the new Center for Research on Contracts and the Structure of the Enterprise at the University of Pittsburgh.

The first section of the issue deals with "Modernizing the SEC." In his paper "Zen and the Art of Securities Regulation," former SEC commissioner Joseph Grundfest identifies "known trends and uncertainties" that make SEC's traditional ideology obsolete. The technological revolution in the fields of computer science and telecommunications, the existence of competing market systems outside SEC's jurisdiction, innovations in financial instruments, increasing globalization of markets and economies, and the rise of

Research in Accounting Regulation, Volume 9, pages 231-234.

institutionalinvestors all point to the need for "a new ideology of regulation that is more sensitive to emerging market realities."

An excerpt of an open letter to President Bush, written by former SEC commissioner Edward Fleishman shortly before his resignation from the SEC in 1992, describes how the SEC could achieve the goal of less burdensome regulation. The changes suggested include greater use of rulemaking for exemptive purposes, a more careful consideration of costs rather than only benefits in rulemaking, greater use of performance standards, and less reliance on command-and-control requirements, greater reliance on market mechanisms as self-regulatory tools, clarification of rules and regulations left intentionally obscure, and a "rigorous streamlining of its multi-thousand-page rulebook."

William Baumol and Burton Malkiel consider the "Redundant Regulation of Foreign Security Trading and U.S. Competitiveness." They suggest that there is little theoretical and empirical support for the claim that U.S. investors investing in foreign companies would benefit from requiring such companies to convert their financial statements into U.S. GAAP. Franklin Edwards indicates in "Listing of Foreign Securities on U.S. Exchanges" that the SEC's claim of U.S. investors being disadvantaged as a result of less extensive disclosures by foreign companies is misguided as it rests on the premise that investors are not willing to pay for the added disclosures of U.S. firms in the form of higher stock prices. If additional disclosures induced by U.S. GAAP is valuable to investors, then U.S. firms should have a lower cost of capital in comparison with their foreign competitors. If, on the other hand, investors do not incrementally value these disclosures, then U.S. disclosure requirements may be unusually burdensome and redundant.

The second set of papers deals with the issue of corporate governance. Ronald Gilson describes "The SEC's Response to the One-Share, One-Vote Controversy." He applauds SEC's rule 19c-4 as a "clever, narrowly focused regulatory initiative" that enables companies to have dual class voting structures, while banning the potentially coercive exchange offers or dual class recapitalizations sometimes used to achieve them. SEC's authority to impose this regulation, was, however invalidated by a court decision, and the matter is now in the hands of Congress. The exchanges failed to voluntarily adopt the rule partly in deference to company managers' inclination to use these recapitalizations as a takeover defense.

Michael Ryngaert suggests a "guiding philosophy" for "An Appropriate Federal Role in the Market for Corporate Control." He feels that Federal policy should "clear the deck of state and federal laws that currently hinder the operation of the takeover market." Federal legislation of a "one-size-fits-all variety" reduces flexibility in investor monitoring and makes it more costly. Hence, managers should be free to propose their own anti-takeover amendments, provided these are subject to shareholder vote.

Bernard Black considers "Next Steps in Corporate Governance Reform: 13(D) Rules and Control Person Liability." He argues that the recently reformed proxy rules instituted by the SEC "make it difficult and legally risky for institutional investors to own large percentage stakes in particular companies, or to play an active role in corporate governance." He focuses on the 13(d) disclosure rules and control person liability rules. The detailed disclosures about ownership and intent that is required under section 13(d) discourage shareholders from owning over 5% of a company's stock, discourage 5% owners from becoming active in corporate governance, and discourage smaller shareholders from acting jointly. Control person liability rules impose unlimited liability on large shareholders for securities laws violations by the companies in which they have invested. Black advocates regulatory change in these areas, as they are "among the major obstacles to institutional oversight."

In "Mutual Funds in the Boardroom," Mark Roe argues that mutual funds may be the best suited of all institutional investors in the United States for playing an active role in corporate governance. He recommends that the current subchapter M tax rules, and certain provisions of the Investment Companies Act of 1940 be modified. Of particular importance are provisions discouraging mutual funds from taking large positions in companies, which have led mutual funds into their current "hyperdiversification."

To cap off the discussion on corporate governance, issues relating to executive compensation are addressed by three articles. First is a documentation of the proceedings of a "Shadow SEC Roundtable on the New Disclosures of Executive Pay." The shadow SEC is a group of prominent financial economists who seek to bring economic analysis to SEC regulatory policy. The discussion centers around the expected costs and benefits of the new disclosures with compensation adviser Jude Rich, and academics Michael Jensen and Kevin Murphy. In the second article, "An Overview of the Executive

Compensation Debate," Greg Jarrell, former SEC Chief Economist, explains his stand in the Roundtable against the new disclosures. He reasons that "there is a fundamental inconsistency between the 'populist' reforms now being proposed in the political sector, and the policy implications of scholarly research." In the third article, "Accounting for Executive Compensation," Ross Watts explains the new accounting treatment of compensation expense and concludes that "the expected FASB proposal for accounting for executive stock options at a minimum will not harm accounting earnings' ability to measure performance and may improve it."

The final section of the issue consists of two papers related to SEC's "Market 2000" study. In "Organization of the Stock Market: Competition or Fragmentation?" Hans Stoll considers the variety of markets in which common stocks can now be traded, and addresses issues relating to the debate between those who consider these trends to indicate dangerous fragmentation, and others who applaud it as a sign of healthy competition. In his view, "it appears more likely that the observed fragmentation of markets reflects competition in the provision of trading services that arise from the development of more efficient trading procedures."

In "Market Transparency: Pros, Cons and Property Rights," Harold Mulherin argues that perfect transparency in markets—the instant availability of prices, volume, and trader identities to all parties at all times—"even if attainable, is not desirable." Competitive pressures will induce exchanges to provide the appropriate level of transparency. He feels that "regulatory oversight of transparency should be limited to the protection of property rights of the products of an exchange and the enforcement of contracts."

Overall, this set of papers provides a student of accounting regulation with economic perspectives on broad trends and issues relating to the regulation of financial markets, by authors who represent the orthodoxy. The treatment of these issues is simple enough to make them accessible to students in masters and undergraduate accountancy programs. References to more rigorous academic papers are provided for the academic interested in regulation. With the preponderance of economic arguments, the viewpoints do not expressly consider social and political forces influencing financial regulation, and are therefore narrowly focused. However, to a student of accounting regulation, it provides a good starting point in an appreciation of the economics of regulation.

Financial Reporting in North America: Highlights of a Joint Study

(Financial Accounting Series No. 144-B,
Financial Accounting Standards Board, December 1994, 32 pp.)

Reviewed by **Kevin F. Brown**

In August of 1992, the Canadian Institute of Chartered Accountants, the Mexican Institute of Public Accountants, and the U.S. Financial Accounting Standards Board formed a Study Group to compare accounting standards, to explore opportunities for harmonizing these standards, and to improve comparability of financial statements among the three countries. The formation of this Study Group took place during the negotiations of the North American Free Trade Agreement (NAFTA) which was ratified in 1993. The purpose of the Financial Accounting Standards Board's report is to provide a brief overview of the Joint Study which resulted from the efforts of the Study Group.

The highlights of the Joint Study begin with a discussion of capital market structures and economic conditions in Canada, Mexico, and the United States. Not surprisingly, the United States dwarfs its neighbors in all of the categories of economic comparison. However, the authors note that the capital markets in the United States and Canada share many similar characteristics, including a high degree of competition, large numbers of suppliers and consumers of capital, and a wide range of debt, equity, and derivative securities. Mexico's capital market, however, does not share these characteristics. Rather, majority shareholders and the Mexican government exert significant influence

Research in Accounting Regulation, Volume 9, pages 235-237.
Copyright © 1995 by JAI Press Inc.
All rights of reproduction in any form reserved.
ISBN: 1-55938-883-8

on the market. In spite of the differences in capital market structures and economic conditions, the authors feel that similar economic trends, which all three of the countries share, will result in users of financial statements demanding similar information.

Next, this report discusses the standard-setting processes and the conceptual frameworks in the three countries. The authoritative, private sector accounting standards boards are the Accounting Standards Board of the Canadian Institute of Chartered Accountants, the Accounting Principles Commission of the Mexican Institute of Public Accountants, and the Financial Accounting Standards Board in the United States. The standard setters follow a similar due process which includes issuance of exposure drafts. All three bodies have conceptual frameworks which the authors note are generally similar. However, one significant difference noted is that, "Mexico's framework explicitly presumes the routine application of inflation accounting in the basic financial statements."

In order to illustrate significant differences in generally accepted accounting principles (GAAP), the Study Group examined the GAAP reconciliations of 176 companies (170 Canadian and 6 Mexican) which filed financial statements with the U.S. Securities and Exchange Commission (SEC). The results of this survey contradict two popular misconceptions about the accounting standards of the three countries. First, the Study Group found a tendency for significant differences between Canadian and U.S. GAAP, which are often thought to be very similar. Second, Canadian GAAP and Mexican GAAP, contrary to popular opinion, are not always less conservative, as measured in terms of both shareholders equity and net income impact, than U.S. GAAP.

The majority of the report discusses the areas of significant differences found by the Study Group's survey. These areas include:

- Effects of changing prices
- Business combinations
- Foreign currency translation
- Income taxes
- Earnings per share
- Retiree medical and life insurance benefits

For each of these areas, the authors provide a brief, but informative, synopsis of the accounting standards followed by Canada, Mexico, and the United States. Also, disclosure examples from surveyed companies are included to illustrate each area of difference.

One potential criticism of this study is its relatively narrow scope. Only Canadian and Mexican companies filing financial statements with the U.S. SEC were surveyed. This limitation may result in not considering significant GAAP differences because they were not captured by the study's sample.

The report concludes with recommendations that the three standard-setting bodies commit to a continuing program of liaison and mutual involvement in standard-setting activities, and "establish a standing committee of representatives from the three counties to initiate and maintain cooperative efforts" (pp. 30, 32). Although these recommendations may seem quite vague, the authors suggest very specific opportunities for implementation, including: undertaking joint projects, considering conformity with other countries when new standards are initiated, participating in the discussion of new standards proposed by others' boards when joint projects are not feasible, reviewing and commenting on the projects of other boards, and developing a program for staff exchange among the boards.

Overall, this report provides an excellent introduction to the work of the Joint Study. Although this presentation format does not allow for the reporting of study results in detail, the report does give its readers insight into the importance and complexity of harmonizing accounting standards. Anyone with an interest in international accounting issues should find this report informative.

Reporting on Environmental Performance
by The Canadian Institute of Chartered Accountants

(The Canadian Institute of Chartered Accountants, 277 Wellington Street West, Toronto, Ontario, 1994; $37.50, Canadian, 183 pp.)

Reviewed by **Susan L. Frazier**

In 1993, The Canadian Institute of Chartered Accountants (CICA) published *Environmental Costs and Liabilities: Accounting and Financial Reporting* which provided suggestions for accounting and reporting on environmental issues. In 1994, the CICA expanded on the topic of "green accounting" by publishing *Reporting on Environmental Performance*. This publication provides guidance to companies on establishing effective environmental strategies and reports.

The CICA established a study group consisting of thirteen representatives from public accounting firms, industry, and regulatory agencies. This group was responsible for formulating a sound framework for companies to follow with regard to environmental reporting. The study group first prepared a "discussion paper," which was reviewed by a combination of 100 organizations and individuals. The strong feedback received enabled the group to prepare a complete guide to environmental reporting.

This research study consists of seven chapters and five appendices. The first two chapters discuss the importance of establishing an environmental performance system and review several environmental strategies used by companies. The following five chapters detail the factors to consider when initiating an environmental program and

Research in Accounting Regulation, Volume 9, pages 239-241.
Copyright © 1995 by JAI Press Inc.
All rights of reproduction in any form reserved.
ISBN: 1-55938-883-8

provide a framework for preparing environmental reports. The appendices list examples of environmental principles and codes, industry environmental performance indicators, and narrative disclosures of environmental performance, as well as provide an outline of environmental reporting award programs.

According to the study group, organizations are facing pressure from stockholders, regulators, customers, suppliers, and other third parties to disclose environmental practices. These stakeholders want to be assured that the environmental practices of the company will not hinder profitability or damage the organization's reputation in the marketplace. In response to these pressures, organizations have begun to establish environmental strategies, that range from reacting to problems as they occur to establishing a comprehensive sustainable development system. Determining the appropriate strategy for each company depends on the organization's objectives and view of its role in society. The exhibits in Appendix A outline examples of environmental principles and codes of practice that will help organizations develop a successful environmental program.

After an organization determines its environmental reporting objectives, a report for internal and external users needs to be created. To help facilitate the development of such a report, the study group established an environmental reporting framework which includes the following major categories: (1) organization profile, (2) environmental policy, objectives and targets , (3) environmental management analysis, and (4) environmental performance analysis. In Appendix D, examples of disclosures of environmental policies, objectives, and results are provided. The study group believes this report format will help companies organize their environmental practices and help stakeholders gain a better understanding of the relationship between the organization and the environment.

Developing environmental performance indicators is another essential to successful environmental reporting. Indicators can be financial or nonfinancial, but should be consistent with the organizations objectives. Appendices B and C contain examples of such indicators for six different industries. Standardizing environmental indicators throughout industries will help the stakeholder benchmark the environmental performance of an organization.

As detailed environmental reporting has become such a major concern of the stakeholders in companies an award system has been devised by Canada and the United Kingdom. This award system was implemented to encourage the environmental reporting process and thus heighten environmental awareness of companies. A description and the judging criteria for each award is provided in Appendix E.

In conclusion, the study group believes that developing an effective environmental report requires a strategy whereby the audience is clearly defined and the environmental indicators meet the needs of the stakeholders. As more companies begin to establish environmental policies, industry indicators will become more standardized. However, without some type of standardized reporting mechanisms, the accounting profession will be unable to fully implement comprehensive environmental reporting standards.

The study group preparing this report did an excellent job at providing a framework for organizations to follow when developing an environmental reporting process. For any organization establishing or expanding their environmental performance reporting, this study would be a very valuable resource.

EDITOR'S NOTES

REPORTING REFORMS IN THE ACCOUNTING PROFESSION:

MARKET TIERS AND CHANGING TRADING RULES MAKE REFORM A NECESSITY

Gary John Previts and Larry M. Parker

ABSTRACT

Global market demand for capital and changes in trading rules for securities have shaped a "private" tier of institutional and sophisticated investors separate from the "public" tier. This helps illustrate the need for the accounting profession to be aware of significant shifts in the way businesses report on operations. Although the profession has changed its role in preparing, attesting, and reporting, preparing for a response to change has not been a hallmark of accountants. Changes such as those recommended by the AICPA Special Committee on Financial Reporting (Jenkins Committee) are important; however, much remains to be done if this report is to affect the information structure of the investment process.

Research in Accounting Regulation, Volume 9, pages 245-250.
Copyright © 1995 by JAI Press Inc.
All rights of reproduction in any form reserved.
ISBN: 1-55938-883-8

245

As the First Century of the CPA profession draws to a close (dating from the passage of the CPA Law in New York State, 1896 to now), some reflection on the role of the CPA, and financial reporting of our evolving instant global capital market, seems warranted.

A century ago, the notion of an investment grade security and massive retail securities trading was only beginning to be envisioned. Gradually, through innovations such as the consolidated holding company, which publicly traded shares, and the investment trust, which debuted in the 1920s, the market expanded to the point where today it seems nearly every working person invests directly or indirectly in capital market instruments—be it simple money market certificates, or derivative securities, or pension or mutual funds and debt instruments—of all kinds. This investment activity, and the corresponding development of information and disclosure rules and policies in the stock exchanges and by government agencies, has—along with the development of income taxation and computer information technology—underwritten the growth of the CPA professional, initially in public practice, and now clearly in the preparation and reporting of information in publicly held operating/ production and investment companies.

As noted in the descriptive paper in this volume by Stephen Young, the demographics of the CPA profession have changed dramatically in recent years. So too have the demographics of investor/ shareholders. By 1990, as discussed by Securities and Exchange Commission Chairman Breeden when he testified before Congress in November 1991, ownership of publicly traded equity instruments by individuals had decreased to less than 50%, compared with 77% in 1955. In another study reported by New York University economist Edward Wolff in 1992, the wealthiest individuals in the land (those in the top half percent of the population) held 29% of all stock and 41% of all bonds in 1989.

About the same time, the SEC, under Chairman Breeden, sponsored trading Rule 144, which became effective in 1990 and which allows expanded exemptions to foreign security issuers and issues of domestic restricted securities (so called letter stock) to trade in a securities market among themselves as "qualified institutional buyers." This permitted, in SEC terms, a "first step toward achieving a more liquid and efficient institutional resale market for unregistered securities." A "private" screen-based computer market, PORTAL (Private Offerings, Resales, and Trading through Automated

Linkages) was established to execute these secondary trades among Qualified Institutional Buyers (QIBs) and other sophisticated investors. To some, the creation of PORTAL, and the dominant presence of institutions in the public capital markets, contributes to the efficiency of the exchange of capital, while promoting placement of much needed U.S. capital to foreign entities, without requiring institutional and sophisticated investors to pay the added cost or "tax" of registration and disclosure which the 1933 and 1934 Securities laws' filings impose for access to public capital markets.

In the public capital markets, it is argued average (unsophisticated) individual investors benefit from the presence of sophisticated and institutional investors, because the former serve as efficient forces in assessing information. As well, these groups are price makers, and, assuming they act rationally and process information efficiently, they may even afford some "price protection" to the small individual investor who is a price taker.

Further, to reduce transparent advantage, or to level the playing field, SEC Rule 144A imposes a test to restrict from PORTAL trading any securities in the same "class" as securities listed on a U.S. stock exchange or an over-the-counter system.

This new architecture for U.S. Capital markets, it is argued, will provide efficiency, fairness, and competitive capital pricing. What occurs to others, who are less convinced of the benefits of such a two-tiered (private and public) market, is the ability of institutions and sophisticated individual investors to be players in investment instruments of the same companies in both markets (albeit not in identical securities). For example, a company which has issued publicly traded common stock may now issue and trade more freely any unregistered stock to private investors, and this seems likely to impact the publicly traded stock.

What exacerbates this uneasiness is recalling the way former SEC Chairman David Ruder, in 1988, characterized the October 1987 "Market Break." Speaking to a group of accounting academics at the Arthur Andersen Education Center near Chicago, he called it a "rush to the exits by institutional investors." There were, as well, many reports of "front running" by institutional traders in public capital markets, where they moved to unwind their institutional positions before they assisted or concerned themselves with individual investor clients.

The question becomes then, "What is the significance of the multi-tier market in bonds and equities?" Writing in *Barron's*, Edward Wyatt (1989) noted "the paucity of information on bond prices and bond market trading that faces the individual investor is nothing new." And, despite frequent inquiries by both the GAO and the SEC, one commentator noted, "frankly, nothing has changed."

So it may be a "fact of investing life" that a two-tiered market (some might argue it is a three-tier market—institutions, sophisticated individuals, and average individuals) does not afford equal access to all investors to the same trading places, and to information about public companies which can be of value in trading all securities (registered and unregistered) of a publicly held company. It may be that the paucity of ongoing academic research about the information needs of various investor/user groups in both bond and equity markets has contributed to the inability to identify or address related disclosure issues with more effectiveness (Robinson 1993, 210-211).

Further, it can be noted that at least since 1982 Regulation SK does require reporting of information about the fact of unregistered securities trades in filings with the SEC under Item 701. Again, without research it is not clear if the average individual investor in public markets is or is not properly informed at least as to the currency/timeliness of such disclosures.

The accounting profession's role in all of this, as the provider/preparer/attestor of information to owners/investors of publicly traded operating companies and investment/mutual companies, is now even more challenging to characterize.

If these new tiers of market trading dislocate information at some levels in favor of traders at another level, what effect does this have on the "price protection" which in the past, it could be argued, was at least available to the individual, small investor, who relied on the price in an efficiently processed information setting?

Federal disclosure laws, etched in the legislation of over 60 years ago, seem suspect in operation, if not in principal. The recent AICPA Special Committee (Jenkins) recommendations proffer to remodel or re-engineer disclosure in an important way. From our vantage, it is a better way to make available in public disclosures the type of business information which analysts and sophisticated investors use. Reforming the content of the "classic" general purpose financial statement/report—to be a more inclusive business report—has, however, encountered the customary opposition from those who employ the "it ain't broke" line of defense.

A well-maintained 1965 model car, now a collectors item, "ain't broke," either. But it is more a museum piece than it is a safe contemporary mode of transportation. So too with the "general purpose" financial report. It needs to be replaced as quickly as the profession, the FASB, and the Commission can reorient their regulatory relationships to accomplish business reporting.

Even with business reporting in place, it will take many years to re-educate academics, preparers, and auditors to form the type of disclosure habits consistent with servicing the needs of users of business reports. This reform addresses only the *operating* company reporting environment, not the disclosures of the investment and mutual fund companies, which are themselves achieving new levels of popular individual investor involvement. Certainly as individual investors begin to evaluate performance and compensation among mutual fund managers, especially given the dismal performance of such funds during 1994, more disclosure of performance bearing information—perhaps using the model of rating agencies—will be needed.

Changing to a global market while maintaining confidence in the variety of instruments and systems is a necessary condition for our private property capital system. Disclosure reform must be a constant, not a variable, in the quest to maintain acceptable fairness in our changing marketplace. Academics who conduct research in our field have limitless opportunities to contribute to the study of market events. This publication will work to be one instrument to encourage and publish reports on such inquiry.

ACKNOWLEDGMENTS

In preparing this perspective, we have benefited from discussions with our colleagues, to include Julia Grant and Robert Bricker, and doctoral students Nandini Chandar and Stephen Young.

REFERENCES

Afterman, A.B. 1995. *SEC Regulation of Public Companies*. Englewood Cliffs, NJ: Prentice-Hall.

American Institute of Certified Public Accountants. 1994. Focus on: Recommendations of the Special Committee on Financial Reporting. *Journal of Accountancy* (October): 41-46.

Block, S. 1992. SEC clears rules to ease big firms raising of money. *The Wall Street Journal*, October 22, p. A13.

CCH Incorporated. 1995. *SEC Handbook*.

Elloitt, R.K. 1994. The future of audits. *Journal of Accountancy* (September): 74-82.

Hicks, J.W. 1993. *Resales of Restricted Securities*. Clark Boardman Callaghan.

Previts, G.J. ed. 1992. Financial reporting in an investor fund economy. In *Research in Accounting Regulation*, Vol. 6, pp. 201-210. Greenwich, CT: JAI Press.

Robinson, T.R. 1993. External demands for earnings management: The association between earnings variability and bond risk premia. Unpublished doctoral dissertation, UMI Dissertation Services.

Torres, C. 1992. Big board set to allow bypass of floor trades. *The Wall Street Journal*, October 26, p. C13.

Wolff, E. 1992. The rich are still getting richer. *U.S. News & World Report*, November 9, p. 16.

Wyatt, E.A. 1989. Trading in the dark: Real bond prices are hard to come by. *Barron's*, November 6, pp. 15, 43, 46.

Advances in Management Accounting

Edited by **Marc J. Epstein,** *Graduate School of Business Stanford University*

Associate Editor: **Kay M. Poston,** *Accounting Faculty, Arizona State* University-West

Volume 4, 1995, 264 pp. $73.25
ISBN 1-55938-882-X

CONTENTS: Introduction: Monitoring Vital Signs of the Financial Health of Suppliers to Jit Systems: Implications for Organizations and Accountants, *Richard J. Palmer, Donald W. Gribbin, and Marvin W. Tucker.* A Case Study of the Organizational Effects of Accounting Information Within a Manufacturing in Environment, *Thomas L. Albright and Thomas A. Lee.* Implementing Activity Costin: The Link Between Individual Motivation and System Design, *Kent L. Henning and Frederick W. Lindahl.* Management Accounting Systems and Contingency Theory: In Search of Effective Systems, *Granger Macy and Vairam Arunachalan.* Finding the Missing Pieces in Japanese Cost Management Systems, *Robin Cooper and Cecily A. Raiborn.* Legacy Costing, Fuzzy Systems Theory, and Environmentally Conscious Manufacturing, *Carol M. Lawrence and Alley C. Butler.* Initial Reponsibility, Prospective Information, and Managers Project Evaluation Decision, *Paul D. Harrison and Adrian Harrell.* The Differential Effect of Information Asymmetry on the Relations Between Budgetary Participation and Departmental Performance, *Alan S. Dunk.* The Effect of Socially Desirable Responding (SDR on the Relation Between Budgetary Participation and Self-Reported Job Performance), *Hossein Nouri, Gary Blau, and Abdus Shahid.* The Effect of Divisional Interdependence on The Use of Outcome-Contingent Compensation, *Paul Kimmel and Leslie Kren.* Work Commitment of Management Accountants: The Development of an Efficient and Reliable Measure, *Dennis M. Bline and Wlda F. Meixner.* An Empirical Evaluation of Student Response Patterns to Introductory Cost Accounting Questions: Evidence from an Expert System, *Paul M. Goldwater and Timonty J. Fogarty*

Also Available:
Volumes 1-3 (1992-1994) $73.25 each

Volume 1, In preparation, Fall 1995
ISBN 1-55938-753-X   Approx. $73.25

CONTENTS: The Concept of Trust and Institutional Development in Auditing, *Steven E. Kaplan and Robert G. Ruland.* The
Relation Between Audit Structure and Public Responsibility: Audit Firms Propensity to Qualify Bankruptcy-Related Opinions, *Joseph V. Carcello, Dana R. Hermanson, and H. Fenwick Huss.*
Commentary on Auditors Public Responsibility, *Jack C. Robert-
son.* The Relation Between Audit Structure and Public Responsibility: Audit Firms Propensity to Qualify Bankruptcy-Related
Opinions: A Comment, *J. Edward Ketz.* Toward an Understanding of the Philosophical Foundations for Ethical Development of
Audit Expert Systems, *Steve G. Sutton, Vicky Arnold, and Thomas D. Arnold.* Comments on "Toward an Understanding of the
Philosophical Foundations for Ethical Development of Audit Ex-
pert Systems", *Jagdish S. Gangolly.* Further Toward an Understanding of the Philosophical Foundations for Ethical
Development of Audit Expert Systems, *Jesse F. Dillard.* On Ethi-
cal Behavior in Social, Political, and Legal Environments, *Steve
G. Sutton, Vicky Arnold, and Thomas D. Arnold.* An International
Comparison of Moral Constructs Underlying Auditors Ethical
Judgments, *Jeffrey Cohen, Laurie Pant, and David Sharp.* An
Analysis of International Comparison of Moral Constructs Under-
lying Auditors Ethical Judgments, *Joseph J. Schultz, Jr.* Com-
ments on an International Comparison of Moral Constructs
Underlying Auditors Ethical Judgments, *W. Morley Lemon.* Do
Expected Audit Procedures Prompt More Ethical Behavior? Evi-
dence on Tax Compliance Rates, *Wanda A. Wallace and Chris-
topher Wolfe.* Discussion Comments on "Do Expected Audit
Procedures Prompt More Ethical Behavior? Evidence on Tax
Compliance Rates", *Mary S. Doucet.* Commentary on the Ethics
of Compliance with Tax Laws and Regulations, *Andrew Baily.* Application of Virtue Ethics Theory: A Lesson from Architecture,
John Dobson and Mary Beth Armstrong. Professional Services
and Ethical Behavior, *Haim Falk.* The Moral Expertise of Auditors: An Exploratory Analysis, *John T. Sweeney.* Ethical Development of Accountants: The Case of Canadian CMAS, *Lois Deane
Etherington and Leah Schulting.* Moral Judgment and Moral Cognition: A Comment, *James Gaa.* Gender Diversity Driven Changes in the Public Accounting Workplace: A Moral Intensity
Analysis, *Karen Hooks and Thomas Tyson.* Ethical Decision
Making in Organizations: A Management-Employee Whistle
Blowing *Model, Don W. Finn.* An Exploratory Study of Accounting
Students Professional Attitudes: Implications for Accounting Ed-
ucation, *Patricia Casey Douglas, Randoph T. Barker, and Bill N.
Schwartz.* Perceptions of Senior Auditors: Ethical Issues and
Factors Affecting Actions, *Elizabeth M. Dreike and Cindy Moeck-
el.* Misleading Annual Report Presentation: Ethical and Financial
Considerations, *Timothy J. Louwers and Robin R. Radtke.*

J
A
I

P
R
E
S
S

J A I P R E S S

Research in Accounting Regulation

Edited by **Gary John Previts**, *Department of Accountancy, Case Western Reserve University.*

Associate Editors: **Larry Parker**, *Case Western Reserve University,* and **Robert K. Eskew**, *Purdue University*

Volume 8, 1994, 263 pp. $73.25
ISBN 1-55938-402-6

CONTENTS: MAIN PAPERS. An Exploratory Content Analysis of Terminology in Public Accounting Firms Responses to AICPA Peer Reviews, *Wanda A. Wallace and Karen S. Cravens.* An Empirical Investigation of Problem Audits, *Bhani Raghunathan, Barry L. Lewis, and John H. Evans, III.* Cost-Benefit Analysis and Accounting Regulation, *Steven Maijoor.* Toward a Global Reporting Model: Culture and Disclosure in Selected Capital Markets, *Robert J. Kirsch.* Consolidation Policies and Procedures Discussion Memorandum: An Examination of the Potential Impact on Reporting Quality, *Robert E. Hoskin and Andrew J. Rosman.* Commentary on Consolidation Policies and Procedures Discussion Memorandum: An Examination of the Potential Impact on Reporting Quality, *Paul Pacter.* The Accounting Thought of Newman T. Halvorson (1908-1992), *Robert Bloom, Marilynn Collins and Jayne Fuglister.* Agency Cost Explanations for the Demand for Differentiated Monitoring Activity, *Michael T. Dugan and Cindy D. Edmonds.* PERSPECTIVES. Slaying the Sacred Cow: Ridding Ourselves of Conservatism, *Donald R. Nicols and Larry M. Parker.* Regulatory Barriers to a Financial Innovation: Single Stock Mutual Funds and Some Related Disclosure Issues, *Arthur J. Wilson and Stephen J. Young.* The Future of Financial Reporting, *Edmund L. Jenkins.* The Noblesse Oblige of Accounting, *Gerhard G. Mueller.* BOOK REVIEWS. The Knowledge of Strategy, *By Nathan Grundstein.* Reviewed by *Walter J. Kennamer.* Capital Ideas: The Improbable Origins of Modern Wall Street, *By Peter L. Bernstein.* Reviewed by *Stephen J. Young.* Cumulative Index: Research in Accouting Regulation, Volumes 1-8 (1987-1994), Prepared by *Sulaiman Al-Tuwaijri.*

Also Available:
Volumes 1-7 (1987-1993) $73.25 each

Advances in Accounting

Edited by **Philip M.J. Reckers,** *School of Accountancy, Arizona State University*

Associate Editors: **Eugene G. Chewning,** *School of Accounting, University of South Carolina,* **Karen Hooks,** *School of Accountancy, Arizona State University,* **Buck K.W. Pei,** *School of Accountancy, Arizona State University* and **Thomas Schaefer,** *Department of* Accounting, *Florida State University*

Volume 13, In preparation, Fall 1995
ISBN 1-55938-881-1   Approx. $73.25

CONTENTS: MAIN ARTICLES. The Determinants of Audit Delay, *Michael Aitken, Freddie Choo, Michael Firth, and Roger Simnett.* The Audit Review Process and Its Effect on Auditor's Assessments of Evidence from Management, *Urton Anderson, Lisa Koonce, and Gary Marchant.* Evidence of the Security Market's Ex Ante Assessment of Differential Management Forcast Accuracy. Assessing Accounting Doctoral Programs By Their Graduates' Research Productivity, *James R. Hasselback and Alan Reinstein.* An Investigation of the Feasibility of Using Statistical-Based Models as Analytical Procedures, *Kenneth S. Lorek, Stephen W. Wheeler, Rhoda C. Icerman, and David Fordham.* An Investigation of the Determinants of Accounting Method Choice Among Initial Public Offering Firms, *John D. Neill, Susan G. Pourciau, and Thomas F. Schaefer.* Examining the Dimensionality of the Ethical Decision Making Process of Certified Management Accountants, *R. Eric Reidenbach, and Donald P. Robin.* Additional Evidence of Auditor Changes: The Effects of Client Financial Condition, *Earl R. Wilson, Inder K. Khurana, and W. David Albrecht.* **PROFESSIONAL NOTES.** On the Use of Contingent Valuation Methods to Estimate Environmental Costs, *Marlbeth Coller and Glenn W. Harrison.* Accounting for Debt Conversions: Current and Future Alternatives, *Alan K. Ortegren and Thomas E. King.* Auditor Materiality Judgement and Consistency Modifications: Further Evidence from SFAS 96, *Michael L. Costigan and Daniel T. Simon.* Indirect Cash Flow Measures Versus Cash Flows Reported Persuant to SFAS No. 96, *Hanan ElSheikha and Paul Munter.* The Relative Importance of Operating Cash Flows and Accrual Income in Explaining Stock Returns: A Cross-Sectional Approach, *Gary R. Freeman and Glen R. Larsen.*

Also Available:
Volumes 1-12 (1984-1994)
 + Supplement 1 (1989) $73.25 each

J A I   P R E S S

Advances in Taxation

Edited by **Thomas M. Porcano,** *Department of Accountancy, Miami University*

Volume 7, 1995, 194 pp. $73.25
ISBN 1-55938-910-9

CONTENTS: Editorial Board. Ad Hoc Reviewers. AIT Statement of Purpose. Editorial Policy and Manuscript Form Guidelines. The Effect of the Tax Reform Act of 1986 on the Composition of Executive Compensation Packages, *Steven Balsam.* An Investigation of Taxpayers Framing Behavior, *Otto H. Chang.* Using Taxpayer Perceptions of Fairness to Redesign the Federal Income Tax Structure, *Phyllis V. Copeland and Philip J. Harmelink.* Diversity in the Professional Tax Preparation Industry and Potential Consequences for Regulation: Linking Attitudes and Behavior, *Andrew D. Cuccia.* Preferences of Small Businesses for Characteristics Important to the Selection of a Tax Preparation Firm, *Peggy A. Hite, Linda M. Plunkett, and Deborah H. Turner.* The Impact of the Omnibus Budget Reconciliation Act of 1987 on Shareholders of Publicly-Traded Partnerships, *Dan L. Schisler and James M. Lukawitz.* Tax Burdens of Large and Small Publicly-Traded Corporations, *Haroldene F. Wunder.*

Also Available:
Volumes 1-6 (1987-1994) $73.25 each

JAI PRESS INC.
55 Old Post Road # 2 - P.O. Box 1678
Greenwich, Connecticut 06836-1678
Tel: (203) 661- 7602   Fax: (203) 661-0792

Printed and bound by CPI Group (UK) Ltd, Croydon, CR0 4YY

08/05/2025

01864950-0003